*Power and pauperism* aims to provide a new perspective on the place of the workhouse in the history and geography of nineteenth-century society and social policy. The workhouse system is set in the wider context of relationships of social and political power in an age associated, paradoxically, with both *laissez-faire* and increasing state regulation. The study pays particular attention to conflicts over Poor Law policy and Workhouse design. Dr Driver demonstrates that despite appearances the workhouse system was far from monolithic, and that official policy was beset with conflict: his study combines a national perspective on the system with a sensitivity to regional variation in policy and practice. The analysis of patterns of relief regulation and institutional provision presented here provides, for the first time, a truly national picture of the workhouse system in operation. *Power and pauperism* emphasises the need to link the study of social policy with wider patterns of power (drawing on much recent social theory), and the value of a geographical perspective in the study of social policy. The book as a whole offers a challenging new interpretation of the historical geography of social policy in nineteenth-century Britain.

*Cambridge Studies in Historical Geography 19*

---

POWER AND PAUPERISM

## *Cambridge Studies in Historical Geography*

*Series editors*
ALAN R. H. BAKER, RICHARD DENNIS, DERYCK HOLDSWORTH

Cambridge Studies in Historical Geography encourages exploration of the philosophies, methodologies and techniques of historical geography and publishes the results of new research within all branches of the subject. It endeavours to secure the marriage of traditional scholarship with innovative approaches to problems and to sources, aiming in this way to provide a focus for the discipline and to contribute towards its development. The series is an international forum for publication in historical geography which also promotes contact with workers in cognate disciplines.

1 Period and place: research methods in historical geography. *Edited by* A.R.H. BAKER *and* M. BILLINGE
2 The historical geography of Scotland since 1707: geographical aspects of modernisation. DAVID TURNOCK
3 Historical understanding in geography: an idealist approach. LEONARD GUELKE
4 English industrial cities of the nineteenth century: a social geography. R. J. DENNIS*
5 Explorations in historical geography: interpretative essays. *Edited by* A. R. H. BAKER *and* DEREK GREGORY
6 The tithe surveys of England and Wales. R. J. P. KAIN *and* H. C. PRINCE
7 Human territoriality: its theory and history. ROBERT DAVID SACK
8 The West Indies: patterns of development, culture and environmental change since 1492. DAVID WATTS*
9 The iconography of landscape: essays in the symbolic representation, design and use of past environments. *Edited by* DENIS COSGROVE *and* STEPHEN DANIELS*
10 Urban historical geography: recent progress in Britain and Germany. *Edited by* DIETRICH DENECKE *and* GARETH SHAW
11 An historical geography of modern Australia: the restive fringe. J. M. POWELL*
12 The sugar cane industry: an historical geography from its origins to 1914. J. H. GALLOWAY
13 Poverty, ethnicity and the American city, 1840–1925: changing conceptions of the slum and ghetto. DAVID WARD*
14 Peasants, politicians and producers: the organisation of agriculture in France since 1918. M. C. CLEARY
15 The underdraining of farmland in England during the nineteenth century. A. D. M. PHILLIPS
16 Migration in Colonial Spanish America. *Edited by* DAVID ROBINSON
17 Urbanising Britain: essays on class and community in the nineteenth century. *Edited by* GERRY KEARNS *and* CHARLES W. J. WITHERS
18 Ideology and landscape in historical perspective: essays on the meanings of some places in the past. *Edited by* ALAN R. H. BAKER *and* GIDEON BIGER
19 Power and pauperism: the workhouse system, 1834–1884. FELIX DRIVER

Titles marked with an asterisk * are available in paperback

# POWER AND PAUPERISM

## The workhouse system, 1834–1884

FELIX DRIVER

*Royal Holloway and Bedford New College*
*University of London*

PUBLISHED BY THE PRESS SYNDICATE OF THE UNIVERSITY OF CAMBRIDGE
The Pitt Building, Trumpington Street, Cambridge, United Kingdom

CAMBRIDGE UNIVERSITY PRESS
The Edinburgh Building, Cambridge CB2 2RU, UK
40 West 20th Street, New York NY 10011–4211, USA
477 Williamstown Road, Port Melbourne, VIC 3207, Australia
Ruiz de Alarcón 13, 28014 Madrid, Spain
Dock House, The Waterfront, Cape Town 8001, South Africa

http://www.cambridge.org



First published 1993
Reprinted 1995
First paperback edition 2004

*A catalogue record for this book is available from the British Library*

*Library of Congress cataloguing in publication data*

Driver, Felix.
Power and pauperism: the workhouse system, 1834–1884 / Felix Driver.
p. cm. – (Cambridge studies in historical geography: 19)
Includes bibliographical references and index.
ISBN 0 521 38151 7 hardback
1. Almshouses – England – History – 19th century. 2. Workhouses – England – History – 19th century. 3. England – Social policy. 4. Public welfare – England – History – 19th century. 5. Poor laws – Great Britain – History – 19th century. I. Title. II. Series.
HV63.G7D75 1993
362.5′83′094209034 – dc20 92–20031 CIP

ISBN 0 521 38151 7 hardback
ISBN 0 521 60747 7 paperback

# Contents

# Figures

# Tables

# Acknowledgements

I have incurred so many debts in the course of researching and writing this book that to enumerate them all would itself take up a small volume. Space permits me to acknowledge just a few of the most important of them. The book arises – if not phoenix-like, at least in terms of lineage – from a doctoral thesis completed at the University of Cambridge. I am pleased to acknowledge here the initial financial support of Cambridge University and the British Academy. I am also indebted to a large number of librarians and archivists, especially Janet Burhouse and her colleagues at Kirklees District Archives, Huddersfield. In the Universities of Cambridge and Exeter, Ian Gulley, Jim Davis and Andrew Teed provided high-quality technical and cartographic support; at Royal Holloway and Bedford New College, Roy Davis and Justin Jacyno helped to prepare illustrations for publication. I have also gained much from friends and colleagues at these three institutions. In particular, I would like to thank Derek Gregory, for his constant encouragement, sound advice and good humour as research supervisor, and Chris Philo, for putting up with stories from the Bastile for so long. I am grateful to Alan Baker, Robin Butlin, Anne Crowther, Richard Dennis and Dorothy Thompson for their help and guidance at key stages of the research. I must also thank the Principal of Royal Holloway and Bedford New College for granting me a period of sabbatical leave during which this book was finished. Julian Jacobs, Anne Witz, Michael Heffernan and David Gilbert provided much-needed moral support at critical moments in what is euphemistically called 'the research process'. Finally I owe Daphne Cotton a great debt, not least for building the desk on which this volume was finally completed. I offer the book as an inadequate recompense, *in lieu*.

The cover illustration is reproduced by permission of Hitchin Museum (North Hertfordshire District Council). Figure 6.1 is reproduced by permission of Leeds District Archives, and Figure 9.1 by permission of the Kirklees District Archives (both West Yorkshire Archive Service). The maps in Figures 3.1–3.4 are compiled and redrawn from E. M. Pierce, *Town-*

*Country relations in England and Wales in the pre-railway age* (unpublished M.A. thesis, University of London, 1957). An earlier version of chapter 5 appeared in the *Journal of Historical Geography*, published by Academic Press.

# Abbreviations

| | |
|---|---|
| Ad & El | Adolphus and Ellis Law Reports |
| CCE | Committee of Council for Education |
| HGM | Huddersfield Board of Guardians Minutes, KDA |
| HO | Home Office Papers, PRO |
| JRUL | John Rylands University Library, Manchester |
| KDA | Kirklees District Archives, Huddersfield |
| LA | Leeds Archives Department, Leeds |
| LGB | Local Government Board |
| MH | Ministry of Health Papers, PRO |
| NAPSS | National Association for the Promotion of Social Science |
| PLB | Poor Law Board |
| PLC | Poor Law Commission |
| PP | Parliamentary Papers |
| PRO | Public Records Office |
| RAWE | Registers of Authorised Workhouse Expenditure |
| RC | Royal Commission |
| SC | Select Committee |
| TM | Tolson Museum, Huddersfield |
| WYRO | West Yorkshire Records Office, Wakefield |

# Introduction

> A foreigner travelling through the country of England, as he enters almost any significant county town, will see somewhere close adjoining it four large buildings:- One of these will look like a large mansion, the second like a house of refuge, the third like a mimic castle or fortress, and the fourth like a factory or huge storehouse. If now, wondering what these buildings are and for what uses they are kept up, he asks for information of one of the natives, he will be answered in order: number one is the lunatic asylum for the poor, supported by the county; number two is the sick hospital, supported by the benevolence of the people of the district; number three is the gaol, supported by the county; and number four is the workhouse, or as the common people call it, 'the bastile', supported by the Union, or a certain number of parishes composing a district.[1]

In 1865, the *Social Science Review* published an unflattering portrait of the Union workhouse, the 'English bastile'.[2] Credited with none of the virtues and all the shortcomings of other institutions, the workhouse was condemned as an irredeemable failure. Its austere, forbidding appearance suggested to this critic only the deterrent logic of a system in which principles of treatment and cure were subordinated to the dictates of economy and discipline. Even the Guardians' board room, the very seat of local authority, was rudely pictured as 'a mixture of an Old Bailey court, a small chapel and a third class railway waiting room'. The disciplinary regime of the 'bastile' simply crushed its unfortunate inmates under a huge apparatus of rules and regulations, constantly reminding them of their powerless, degraded status as paupers.[3]

This book is concerned with the discourses and practices which helped to create these lasting images of workhouse life. The 'bastile' has gained a quite dismal reputation amongst historians as much as contemporaries. In contrast to the prison or the hospital, for example, it has had few champions and many enemies. For its critics amongst Victorian social scientists, the workhouse was too monolithic, too undiscriminating, to serve a useful purpose within a scientific or humane social policy. The complaints of these specialists were more than matched by popular attacks on the workhouse. 'In the bastile', claimed one Chartist in 1837, one found 'the most virtuous people crowded with the most vicious people on earth, and the treatment of one the same as the treatment of the other, and both worse than the common felon'.[4] Such rhetoric fixed the Union workhouse in the popular imagination as a place of unparalleled dread, unique amongst welfare institutions. In general, historians have tended to draw a similar contrast between the cumbersome bureaucratic rituals of the Union workhouse and the scientific, apparently more 'progressive' regimes of other institutions like asylums or reformatories, many of which have more or less direct descendants in modern institutional forms. The Union workhouse is thus remembered as an institution which 'failed'; an odd detour on the road towards a rational social policy. Yet this verdict is surely a partial one; indeed, in its most unhistorical forms, it merely reproduces the claims of critics, without sufficiently acknowledging the 'wider pattern of incarceration'[5] to which the workhouse belongs. In situating the history of workhouse policy within this wider context, it is necessary to look beyond the walls of the workhouse itself, at a range of other institutions, authorities, pressure groups and activists. For it is only by looking beyond the Poor Law that one can specify more accurately the specific characteristics of the 'bastile' and the features it shared with the other institutions of its time.

In this book, the Union workhouse is situated at the intersection of two histories, those of modern government and institutional discipline. This dual concern with government and discipline is reflected in the use of the term workhouse *system* throughout this book. The bureaucratic context of institutional policy is of particular importance for the history of the workhouse, perhaps even more so than for other institutions. The association of the 'bastile' with officialdom seems to have played a very important part in the negative popular image of the workhouse which survives to this day. This is not to suggest that the origins of and influences on workhouse policy can be explained in terms of some undifferentiated, monolithic process of 'government growth'. Despite the hopes of the architects of the new Poor Law in 1834, the subsequent history of workhouse administration bears the imprint of all kinds of conflicts, struggles and compromises. The mass anti-Poor Law protests of the 1830s and the stubborn recalcitrance of local authorities both indicate, in different ways, that the history of social policy cannot be written in terms of central programmes alone. Officials of central government constantly found their policies modified, frustrated and diverted at local level; such obstructions must therefore find their place in the account which follows.

It should be emphasised here that I am not attempting to write a history of the workhouse system from the point of view of the paupers who experienced it. Rather, I am exploring precisely those features of the system which in a sense made their voices irrelevant to central policy. It is indeed a sad irony that a system which gave rise to a massive archive of paperwork, of books, accounts, correspondence, registers, files, plans, inquiries – in fact an administrative discourse of massive dimensions – should be so obstinately silent on the views and experiences of paupers themselves. For the workhouse system was clearly not designed for paupers; it was rather designed for their management. The establishment was not so much 'baffled by its own paupers'[6], as one Victorian socialist put it, as unmoved by the tales they had to tell. That is not to say that the views and opinions of workhouse inmates went wholly unrecorded; only that they survive in fragmentary form, inevitably marked by the bureaucratic rituals of the system. Take, for example, the case of the Poor Law inspector who took the unusual step of including extracts from graffiti on the walls of vagrant wards in an official report. These messages recorded the movements of characters like 'Saucy Harry and his Moll', 'Bristol Jack', 'The Flying Dutchman', 'Cockney Harry' and 'Dublin Dick', to name but a few of the more colourful examples. The inspector, predictably enough, read them as signs of the immorality and indiscipline of 'professional' tramps, 'confirmation [of the] character and habits of the vagrant class'.[7] Reading such reports from the perspective of the late twentieth century, one can only wonder at this faith in the transparency of the written record. In this case, the writing on the wall was clearly regarded as more reliable than the testimony of the vagrants themselves. These traces of vagrant life may of course take on a rather different meaning in the hands of the cultural historian. Nevertheless, it remains an important task to understand the sources and consequences of the official readings themselves.

The history of the workhouse system is full of these haunting images, traces of lives which officials tried to organise but so often failed to understand. The photograph on the cover of this book, for example, shows a regimented group of pauper children outside Hitchin workhouse, circa 1880. They stand outside the imposing entrance, wearing standard uniforms and caps, passive under the lens of the camera and the eyes of workhouse officers. The presence of a workhouse band gives one indication of official efforts to inculcate in the children a sense of discipline and organisation, anticipating the part they were intended to play in society at large. Government inspectors sent regular reports to London on the spirit and tunefulness of such bands, assuming they were good indicators of the effectiveness of workhouse discipline. (These inspectors were clearly discriminating in their musical tastes; reporting on a pauper band at Leeds, one complained that 'the drums, whenever they played, were used simply to make an overpowering noise'.[8]) The workhouse visiting hours are just visible on the window panes of the building behind, another sign of the orderly, timetabled rhythms of an institutional regime. The portrait as a whole appears to substantiate a common image of life in the workhouse, a drab and monotonous existence ruled by the clock. (The critic of the 'English bastile' cited above describes workhouse children as destitute even of hope: 'They neither laugh as ordinary free children do, nor move like them; when they laugh they tremble, when they run they shuffle, and when they come in obedience to a call they cringe'.[9]) Yet there is something about the image at hand which leaves a more ambiguous impression. Perhaps it is the half-smile on the face of one of the officials (probably a teacher or industrial trainer) standing in the workhouse doorway; or the sullen expression of the lone girl at the front (possibly the daughter of the workhouse master).[10] Or perhaps it is the glance of the child sitting at the end of the front row, a movement which leaves two faces on the print, one obediently submitting to the camera, the other distracted by something beyond its view. We might read this gesture as a sign of a broader truth – that disciplinary regimes do not automatically produce disciplined minds and bodies. The photograph presents us with a regiment of docile bodies; but it can only hint at the possibility that discipline might fail, that surveillance could be subverted.

This book explores the changing nature of workhouse policy and practice in England and Wales during the fifty years which followed the passing of the 1834 Poor Law. Chapter 1 offers a general perspective on what I call the historical geography of social policy, providing a broad framework for the study that follows. It considers two aspects of modern discourses of social policy in particular: the problem of government (especially the administration of territory) and the problem of institutional design (especially the reform of individual conduct). The remainder of the book is devoted to the workhouse system itself. Inevitably, the 1834 Poor Law reform plays a central, though not exclusive, role throughout. The great paradox of modern Poor Law historiography is that the 1834 reform has been seen *both* as the founding moment of the modern welfare state *and* as the symbol of a new ethos of *laissez-faire*. In chapter 2, these themes are addressed in the wider context of state formation in mid-Victorian Britain. This context provides the basis for understanding the common associations between the Union workhouse and particular forms of bureaucratic regulation. Chapter 3 examines the geography of Poor Law government. The transformation of the administrative map under the new Poor Law, though strategically incomplete, had important implications for the character of the new workhouse system. This chapter also explores the geography of relief regulation after 1834, indicating the extent to which central policies were thwarted by, amongst other things, local resistance.

The following three chapters are all concerned with the discourses and practices of the new workhouse system itself. When placed on the map, the 'bastile' appears to be neither myth nor monolith. Chapter 4 outlines the policy context, emphasising the combination of strategies embedded in the 1834 reform itself, especially those concerned with classification, a concept which had both semiotic and disciplinary functions. (These terms are introduced in chapter 1.) Conflicts over the interpretation of classification loom large in the history of official policy. The focus on the 1860s as a key moment in this history is explored further in the analysis of patterns of workhouse building which follows. Chapter 5 provides a unique national picture of the historical geography of workhouse provision during the first fifty years of the new Poor Law. It draws upon a neglected source in an attempt to overcome the dualism between 'local' and 'national' studies which has unfortunately become entrenched in much Poor Law history. Chapter 6 considers debates over workhouse policy and practice in relation to two key groups of paupers: children and the insane. Here the concept of classification (and its corollary, spatial separation) was of central importance. Hence the significance of debates over workhouse design – the layout of the various wards, the height and thickness of the walls, the transparency of the windows. Yet the authorities did not speak with one voice on such matters; as this chapter shows, official policy remained fractured and often contradictory.

Chapter 7 considers the popular politics of the new Poor Law in the industrial North, a region which provided the focus for a sustained campaign against the new system in 1837–1839. This conflict is interpreted in directly territorial terms as a struggle for space – for such things as access to meetings, control over the workhouse, local autonomy. The tactics of the anti-Poor Law movement and the strategies of the central authorities are mapped onto the wider politics of territory shaping popular radicalism during the late 1830s. Particular attention is devoted to the case of Huddersfield, as it was the epicentre of anti-Poor law politics for much of this period. Huddersfield also provides the focus for the following two chapters on local Poor Law policy and practice. These chapters demonstrate both the continuing significance of local resistance to central policy and the unevenness of the project of centralisation formulated in 1834. The historical geography of local administration and institutional policy-making presented here draws the themes of the earlier chapters into a local context, indicating the extent of the gap between official intention and local outcome. Eventually, the shape of workhouse policy did change even in Huddersfield, though this was not the result merely of central directives. The workhouse system in 1884, both in aggregate and in its various parts, bore the imprint of a revolution in the government of pauperism. This book is about the geography of that revolution.

# 1

# Policing society: government, discipline and social policy

There are many ways of writing the history of social policy. One popular text on the British welfare state enumerates seven distinct perspectives on its history, from the Whig to the conspiratorial.[1] Others might revise and extend this list still further; indeed, many of the assumptions implicit in conventional histories of the 'welfare state' have themselves been challenged.[2] This plurality of perspectives suggests a general lack of consensus as to the functions and meanings of social policy. The Fabian and liberal orthodoxies of earlier generations of historians and sociologists have generally been found wanting. During the 1970s, critical theorists focussed their attention on the relationships between social policy and capitalism.[3] In more recent years, simple equations between the development of the welfare state and the interests of capital have been eschewed, and critical writing on social policy has generally abandoned the excessive functionalism of earlier accounts. For some, this has required much more attention to the language of social policy, to the discourses which frame and give meaning to particular policy instruments. For others, it has meant giving more space to the struggles which continually take place over welfare policy and welfare rights. While there remains a wide variety of approaches on offer, what is striking is the extent to which recent approaches focus on the ways in which power relations cross-cut the domain of social policy. These relations are not confined to the economy or society at large, as if social policy merely reflected (or corrected) wider patterns of inequality. Rather, the discourses and practices of social policy are themselves deemed to constitute a field of power relations and inequalities. In very general terms, this is the perspective adopted in this book.

An analysis of these discourses of 'social policy' must begin with the term itself and what it presumes. The concept of social policy assumes the existence of discrete social spaces, bounded in some way by recognisable territorial limits, containing populations which are the objects of 'policy'. Social policy also requires there to be some notion of what is to be policed, and therefore some idea of norms, patterns of conduct, health and welfare. This in turn implies the desire to collect information about the population, on the one hand, and the capacity to issue regulations, on the other; particular forms of knowledge (statistics, the science of states) coupled with particular mechanisms of regulation.[4] In a very general sense, therefore, the emergence of social policy may be associated with the development of modernity and modern states. (As one historian has observed, 'some modern nations have been police states; all, however, are policed societies'.[5]) As an illustration, we might briefly consider the history of the Enlightenment discourse of 'medical police', and its concerns with the health and welfare of the population.[6] Medical police evolved from a developing science of police (*polizeiwissenschaft*) established during the seventeenth and eighteenth centuries, largely but not exclusively in continental Europe. This science of police was increasingly concerned with how to achieve the social goals of security (the public health) and wealth (the public good).[7] It promoted the collection of statistics concerning the condition of the population; in the specific case of medical police, these were designed to facilitate the administrative and legal monitoring of activity influencing the public health. The development of medical police thus signalled the birth of the concept of population as an object of administrative surveillance (and the emergence of what Foucault calls 'biopolitics').[8] In the original sense of the term, 'police' encompassed a wider range of activity than its current use suggests; as well as the prevention of danger (the 'keeping of the peace'), it had the more positive function of the promotion of health and wealth. The etymological connections between police and policy (the words police, policy and politics derive from the Greek terms *polis* and *politeia*) thus signal wider historical associations.

Consideration of these eighteenth-century discourses of police thus opens up some interesting questions about the genealogy of social policy in modern Europe. It might be argued, however, that the idea of 'police' was too closely associated with absolutism and modern versions of mercantilism to take root in Britain.[9] Certainly, it is important to recognise the considerable variations in the concept and role of the state which developed in different parts of Europe. However, it is clear that English and Scottish political theory was in fact less insulated from continental traditions than is sometimes assumed. The writings of Jeremy Bentham, for example, are full of ideas about police and security; indeed, historians of social policy and law enforcement in eighteenth- and nineteenth-century Britain have frequently drawn on Benthamite notions of police in their interpretations of administrative change.[10] Police, in Bentham's hands, is not a discipline of arbitrary power; in fact, it presumes the diffusion of power rather than its concentration. In this respect, one might perhaps distinguish the Benthamite notion of police (which was also shared by Colquhoun and Chadwick, amongst others) from its predecessors. Mitchell Dean thus suggests that the 'police' of liberalism, with its emphasis on rational, calculating individuals and 'natural' economic laws, differed significantly from the 'police' of the mercantilist state.[11] (Malthus is a case in point: 'When nature will govern and punish for us, it is a very miserable ambition to wish to snatch the rod from her hands and draw upon ourselves the odium of the executioner'.[12]) This argument places particular emphasis on the significance of liberalism for modern modes of government, something which is essential if the 1834 Poor Law reform is to be properly understood, as we shall see in chapter 2. More generally, it indicates that the idea of 'police' itself has an historical geography. Clearly, the genealogy of social policy does not consist in simple, unadulterated continuities.

The remainder of this chapter explores two aspects of modern discourses of social policy, focussing on the problem of government and the problem of discipline. Together, these themes provide a general framework for the account of the workhouse system which follows.

## Administrative landscapes: the government of space

There is [from the seventeenth century] an entire series of utopias or projects for governing territory that developed on the premise that a state is like a large city; the capital is like its main square; the roads are like its streets. A state will be well organised when a system of policing as tight and efficient as that of the cities extends over the entire territory. (Michel Foucault, 1982)[13]

To govern is to administer a population within a certain territory. Under this general definition, the art of government is not the sole preserve of states; in modern societies, however, it increasingly becomes so. As Michael Mann has argued, state power is distinguished by its 'unique ability to provide a territorially-centralised form of organisation'.[14] States may thus be differentiated according to the extent to which they routinely penetrate or coordinate civil society; in other words, according to the extent of their 'infrastructural' powers.[15] In a similar vein, Anthony Giddens has recently distinguished between traditional, 'segmental', societies, in which states have a comparatively low administrative 'reach', and modern societies in which the state's infrastructural power is consolidated and centralised. The modern state, he argues, is an effective 'power container', organised by and through territorial markers of various kinds.[16] The state, so to speak, colonises the social, managing and controlling its development. Or as Mann puts it, 'the infrastructural power of the state tends to focus the relations and struggles of civil society onto the territorial plane of the state, consolidating social interaction over that terrain'.[17] Rather than operating over and above society, literally and metaphorically at a distance, the state helps to make modern society what it is.

The work of historical sociologists such as Mann and Giddens raises important questions about the process of state formation.[18] The intensification of the state's infrastructural powers depends to a large extent on its capacity to store information in particular places and to communicate messages, personnel and resources across space. Thus technological innovations associated with writing, printing and transportation are of considerable importance for processes of state formation. It is significant, in this respect, that questions of spatial integration loom large in many historical accounts of the reform of policing, criminal justice and public order in the more specific context of eighteenth- and nineteenth-century Britain.[19] This suggests that an 'infrastructural' perspective has much to offer the historian of social policy. Yet it would be wrong to portray the extension of the infrastructural capacities of the central state as a self-propelling process of rationalisation. As Miles Ogborn points out in a recent study of local power and state regulation in nineteenth-century Britain, the extension of administrative power is always a project, never a certainty; there is always a gap between central policies and local realities. Thus while the territory of the modern state is frequently projected as even and homogenous (as Mann suggests), it often appears in more concrete terms to be fractured and contested.[20] This image of differentiation raises important questions about the geography of the state and what has been called the 'spatial division of power'.[21] To some extent, these questions are addressed in the historical literature on 'central–local relations' (see chapter 2). However, as Ogborn points out, it is misleading to portray these relations in terms of a zero-sum power game, as if the 'centre' and the 'locality' were unchanging, fixed entities, locked in a battle for power. During the mid-nineteenth century, as we shall see, proponents of centralisation envisaged a qualitative change in the relations between the 'centre' and the 'localities'; they hoped not simply to shift the balance of power, but to reconstitute the very terms on which central–local differences were played out.

A purely 'infrastructural' perspective on centralisation requires still further qualification, to the extent that it pays insufficient attention to questions of culture, ideology and meaning. In this context, it is interesting to note that whereas Mann considers the extension of infrastructural power to be perhaps the key distinguishing feature of the modern state, Giddens prefers to speak of surveillance. Here, surveillance is understood as a territorially-based process of administration, designed to regulate the conduct of human beings. According to Giddens, it requires not just the collection and storage of information, but also the routine monitoring of conduct and compliance.[22] Different organisations, including states, have different capacities for surveillance, depending on the various resources at their disposal; again, the ability to store information and communicate messages over space is of critical importance. However, the notion of surveillance involves more than simply administrative efficiency, narrowly conceived; the regulation of conduct, in Giddens' account, depends on normative judgements about deviance and the possibility of reformation. The idea of routine monitoring thus suggests new, more calculable approaches to the regulation of human behaviour. Surveillance, in other words, is not merely about the government of space; it also requires new forms of knowledge, new 'techniques for the disciplining of human difference'.[23] The modern notion of surveillance thus implies new ways of thinking about the possibility of moral regulation. It is to these new ways of thinking that we now turn.

## Disciplinary spaces: design and reform

Morals reformed – health preserved – industry invigorated – instruction diffused – public burthen lightened – economy seated as it were upon a rock – the Gordian knot of the Poor Laws not cut but untied – all by a simple idea in Architecture! (Jeremy Bentham on the Panopticon, 1791)[24]

The surveillance of individual conduct was central to various projects of moral regulation emerging during the eighteenth and nineteenth centuries. The term 'moral regulation' is useful in the present context, because it highlights the connection between the art of government and the regulation of individual conduct.[25] The reformation of 'morals' was arguably the key problem for social policy and social science during this period. The aim was to re-form the undisciplined impulses of paupers, criminals and delinquents, cultivating within them a sense of moral agency.[26] It is important to recognise that the moralising tone of these discourses did not preclude a generalising, even abstract, approach. On the contrary, thinking about morals, about patterns of conduct and obligations, was a way of thinking about society at large; a pattern of thought which was to reach the height of its influence in mid-Victorian England.[27] The science of 'moral statistics', for example, was concerned with the norms, habits and conduct of particular sectors of the population; it mapped the distribution of phenomena such as crime, delinquency, pauperism, disease, drunkenness, insanity and prostitution.[28] The newly-diagnosed condition of 'moral insanity', to take another example, was defined in 1835 as one in which 'the power of self-government is lost or greatly impaired; and the individual is found to be incapable . . . of conducting himself with decency and propriety in the business of life'.[29] The term 'moral' here refers, as in the case of 'moral statistics', to norms or rules of conduct; moral insanity was thus indicated by anti-social behaviour rather than any distortion of the mental faculties. The function of 'moral treatment' was similarly to reform individual conduct rather than to restore the intellect. At the celebrated York Retreat, for example, moral treatment proceeded not by mental therapy but through a regime of 'normal' habits and routines.[30] To be 'demoralised' was, in this discourse, to be beyond the social pale, without discipline.[31] The aim of 'moral' regulation, then, was to improve the *morale*, the self-discipline, of individual human agents.

It was in the sphere of institutional design that the strategy of moral regulation was developed with most effect. In recent years, historians of the prison, the asylum, the hospital and the reformatory during the eighteenth and nineteenth centuries have demonstrated the profound extent to which design itself was conceived as a mechanism of moral discipline. The very terms used in recent historical accounts – the 'moral architecture' of the asylum, the 'moral geometry' of the prison, the 'medico-moral space' of the reformatory, the 'moral universe' of the hospital, the 'school as (moral) machine'[32] – suggests that institutional design had moral functions. In his book *Discipline and punish*, Foucault portrays the birth of the modern prison as emblematic of this radically new form of reformatory power. In place of the scaffold, which exacted punishment in a very public, violent and in some respects inefficient way, Foucault argues, the prison was designed to discipline individual criminals in the privacy of their secluded cells. Within the institution, an intense reformatory regime would mark out the movements of every inmate. The timetables, classifications, dietaries and drills of prison life together orchestrated a *dressage* of bodies; as Foucault puts it, 'discipline fixes, it arrests or regulates movements; it clears up confusion; ... it establishes calculated distributions'.[33] Foucault portrays Bentham's Panopticon design as the exemplar of this mode of disciplinary surveillance. Intended to serve equally as a prison, a workhouse, a factory, an asylum, a school – in fact, any kind of structure in which large numbers of inmates were to be managed – the Panopticon was planned as a circular building, with cells around its circumference and an inspector-governor at its centre (Figure 1.1). An ingenious system of blinds ensured that while the governor could inspect each inmate at will, they would not be able to see whether they were being watched. The light of inspection – the master's eye – would thus penetrate the walls of every cell and every individual conscience. It is in this respect that the Panopticon may be regarded as a paradigm of what Robin Evans has called 'the architecture of moral purpose'.[34]

Foucault's account of the Panopticon has attracted a great deal of criticism.[35] One question arises from the evident differences between institutions like schools, hospitals, asylums and prisons: how useful is it to speak of them simply in terms of a common reformatory model? Foucault himself anticipates this point in *Discipline and punish*: 'There can be no question here of writing the history of all the different disciplinary institutions, with all their individual differences. I simply intend to map on a series of examples some of the essential techniques that most easily spread from one to another'.[36] A rather different criticism emphasises the lack of correspondence between the disciplinary order of institutions and that of society at large. Giddens, for example, argues that the Panopticon cannot 'express' the general nature of

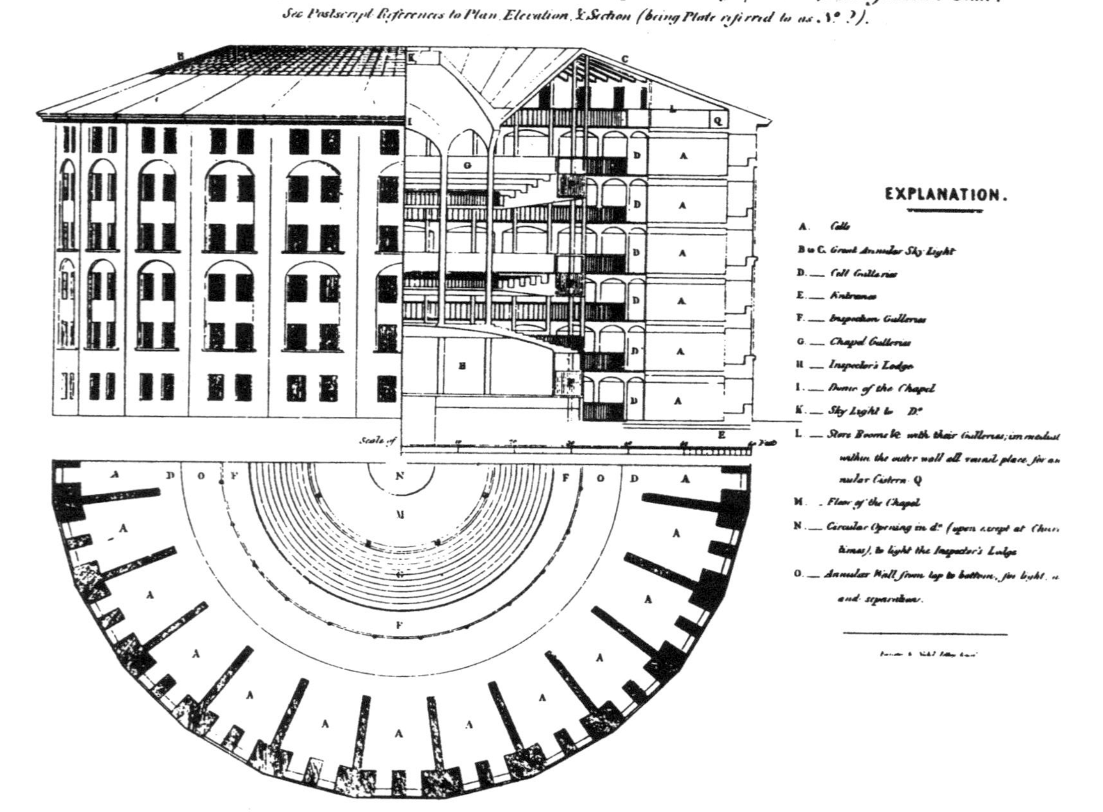


Figure 1.1 Bentham's Panopticon (J. Bowring, *The works of Jeremy Bentham*, 1843)

disciplinary power because total institutions are very distinct kinds of spaces with very specific characteristics.[37] This argument might be extended still further, since the Panopticon is clearly inadequate as an 'expression' of power relations in total institutions themselves. Erving Goffman (the main source for Giddens' ideas about total institutions) devotes a considerable amount of space in his key work *Asylums* to the failures of surveillance in total institutions, discussing the various strategies which inmates and staff devise in order to escape the eye of regulation.[38] It is thus clear that, as an 'expression' of the totality of power relations beyond or within total institutions, the Panopticon model must fail. At the same time, in Foucault's defence, it should be acknowledged that the model was never intended to be read in this way; rather than expressing or reflecting power relations in their entirety, it is presented as the paradigm of a set of institutional programmes.[39] Foucault thus describes the Panopticon as a 'diagram of a mechanism of power reduced to its ideal form . . ., a pure architectural and optical system [abstracted from any] obstacle, resistance or friction'.[40] In fact, *Discipline and punish* contains little concrete discussion of the ways in which such models were put into practice, diffused and resisted. This is an important point because, as Foucault himself remarks, there is a significant difference between the idea of a 'disciplinary' society (in which such programmes proliferate) and a completely 'disciplined' society (in which such programmes are in a sense unnecessary).[41]

It thus becomes important to consider the means by which notions of surveillance and discipline were extended beyond the walls of institutions. Samuel Tuke, the asylum reformer, once suggested that the 'regulations of an asylum should establish a species of espionage terminating in the PUBLIC'.[42] The Secretary of the influential Social Science Association was to go even further, claiming (in 1857) that 'there is nothing in a reformatory school which might not, with proper appliances, be effected for society at large'.[43] The task of building these 'appliances' occupied a veritable army of moral engineers during the nineteenth century, from landscape architects to sanitary scientists.[44] Their labours enabled a reformulation of the very idea of design. The historian James Schmiechen suggests that early nineteenth-century theories of design incorporated two basic principles, which he labels 'associational functionalism' and 'spatial functionalism'.[45] The former, he argues, invoked Enlightenment aesthetic and psychological theories (such as those of Hume and Hartley) which emphasised the relationships between sensation, perception and behaviour. The latter encompassed the notion that moral improvement and social control could be achieved through the manipulation of space. These ideas were promoted by a wide variety of social and sanitary reformers during the late eighteenth and early nineteenth centuries, including such influential figures as Jeremy Bentham (whose Panopticon scheme clearly presumes both kinds of functionalism), John Howard (who might best be described as the patron saint of institutional reformers), John Loudon (designer of houses and gardens) and George Godwin (editor of *The Builder*). While these reformers addressed a wide range of different problems, what united them was their common interest in the links between environmental design and moral reformation. This concern was shared by many of those operating within the new structures of government established during the 1830s and 1840s. It meant that no detail of design, however small, could be ignored; to each environment there was a corresponding form of life. In his comments on plans for a provincial model lodging house submitted to the General Board of Health in 1853, one official thus emphasised the moral significance of the smallest detail:

> With a separate [washing] apparatus in each room, every tenant becomes directly responsible for his own acts. The decent and cleanly will appreciate the accommodation and feel pride in its proper use and good condition, while the dirty and disorderly may by teaching and warning be brought to a higher order of feeling of appreciation of cleanliness and order which, once attained, will never desert him.[46]

This unbounded faith in the moral powers of design was shared by a generation of architects, engineers and administrators.

Contemporary theorists of moral regulation thus placed considerable emphasis on questions of spatial organisation. Using Schmiechen's terms, the principle of 'associational functionalism' represented space as a network of signs which communicated in ever more detailed ways the idea of moral improvement.[47] (More specifically, associationist psychology provided a materialist framework for understanding the process by which environmental stimuli, physical sensations and human perception might be linked.[48]) The principle of 'spatial functionalism' represented space as a mechanical system designed to produce moral subjects. These two principles, and their spatial logics, are also deployed in Foucault's *Discipline and punish*. The idea of associational functionalism, for example, plays an important role in Foucault's account of the 'semio-techniques' of Enlightenment theories of punishment, which proposed an economy of signs centred on the rational, calculating individual. The idea of spatial functionalism is similarly mirrored in Foucault's account of the 'disciplinary' techniques of the prison, which give rise to a new politics of the body.[49] Foucault tends to argue that the semio-techniques were incorporated within, or superseded by, the disciplines, with the Panopticon operating as a kind of bridgehead between the two. However, this image of incorporation presents a rather one-sided account of the genealogy of modern powers of punishment, particularly in the British case, where notions of individualism and the sovereignty of law have been particularly significant. In this context, we might also note that Foucault has been criticised for neglecting the continuing significance of religious frames of reference in the discourse of moral reformation; Ignatieff's model of penal reform, for example, stresses the significance of 'symbolic persuasion' as much as 'disciplinary routinization'.[50] In more general terms, there is a lasting tension within nineteenth-century discourses of social policy, as Mitchell Dean points out, between 'merely semio-technical measures, which assume the problem of self-responsibility to be a question of overcoming artificial obstacles and replacing them with the appropriate signs to guide the choices of the poor' and 'more positive measures which seek to create interventions which actively shape the conditions under which choice is made'.[51] I shall argue in subsequent chapters that this tension was particularly clear in the post-1834 workhouse system. On the one hand, the new Poor Law employed a kind of 'semio-technique', based on the deterrent logic of less eligibility and the figure of the 'independent labourer'. On the other, the new workhouses were much more than simply signs; the institutional spaces themselves were designed to perform social functions.

## Towards a historical geography of social policy

Draw a map, but according to the lines of the age, without levelling and superimpositions. And at least pose the question: what clashes of war, what struggles have left their mark on the paper, what forces keep changing and redrawing the contours? (Dario Melossi, 1981)[52]

An historical perspective on social policy throws light on the provenance of distinctively modern ideas and practices concerning the health and welfare of the population. In this context, the work of Michel Foucault and his associates is of particular interest, not least because there is a particularly radical form of historical interpretation implicit in his genealogical method; history as a defence against essentialist accounts of such things as madness, disease, poverty and crime. As it has become somewhat fashionable to decry Foucault's influence on historical research, we might pause here to consider further the value of his work from the point of view of historians of social policy.[53] For Foucault's emphasis on discourse as an object of inquiry allows us to ask questions not just about experiences and events, but also about the way problems are framed as problems, the way social policies are located in relation to economic and political institutions, the way expertise and authority are constructed, and so on. His focus on the realm of practices, in our case on what Foucault would call the technologies of social policing, suggests new questions for historians of governance and bureaucracy; these are as much 'how' questions as 'why' questions. Finally, his constant recourse to the language of boundaries, locations, separations and colonisations highlights the relative neglect of questions of space and spatial strategies in the history of social policy. In many respects, these emphases are not unique to Foucault; they are also to be found, for example, in some recent work in the historical geography of social policy.[54]

Despite this evident debt, or perhaps because of it, this book nevertheless departs both from Foucault's method and from his particular concerns. Foucault describes *Discipline and punish* as a chapter in the history of punitive reason, a genealogy of the modern disciplinary calculus, rather than a conventional history of prison administration. For this reason, he was not particularly concerned with tracing the various ways in which disciplinary strategies were actually diffused and/or deflected in eighteenth- and nineteenth-century societies. Yet if we are to locate disciplinary programmes in a wider field of power relationships and resistances, historically and geographically, then further questions and other methods are necessary. Two issues are particularly important in the context of the study which follows. First, we need to know more about the actual workings of different institutional regimes and the complex relationships between institutions and the communities in which they are situated.[55] How were institutional programmes actually put into effect? By what means were the practices of institutional discipline implemented or not implemented? To what extent was there an integrated system of institutions? This raises a second issue: the changing nature of state regulation, a subject about which Foucault says surprisingly little in *Discipline and punish*.[56] In this context, what is perhaps most striking is the fractured, dispersed nature of governance in eighteenth- and nineteenth-century societies; as one historian of educational policy put it, 'even during the "revolution in government" of the 1830s and 1840s, polycentricity of power and its dispersal through the provinces remained English characteristics'.[57] In the case of the workhouse system, a focus on the geography of state regulation helps to illuminate the context and consequences of Poor Law reform in 1834. Yet such a perspective remains underdeveloped, because historians have largely failed to put social policy on the map.

The problems of government and discipline associated with modern discourses of social policy have important geographical dimensions. In this chapter, these problems have been framed in explicitly spatial terms, in order to highlight the relative neglect of these issues in existing accounts. This neglect is exemplified in the common dualism between 'local histories', which tend to treat individual places in isolation from any more general picture, and 'national histories', which substitute a focus on central policy and legislation for a truly national perspective. Turning to more theoretically-sophisticated accounts of the sociology of deviance and social control, we find that studies of prisons, police departments and hospitals frequently lack an explicit spatial dimension; the problem, as one sympathetic critic puts it, 'is not so much that these agencies often have no history; they also have little sense of place. They need locating in the physical space of the city, but more important in the overall social space'.[58] In this spirit, another historian has suggested that the construction of an 'historical and geographical map of institutions of social control [would] throw into sharp relief the discourses which came to surround the institutions'.[59] Such arguments indicate that the idea of an historical geography of social policy is not only possible but necessary. By placing the workhouse system on the map, as it were, we not only create an historical geography; we also write better history.

# 2

# Social policy, liberalism and the mid-Victorian state

The debate which preceded the reform of the Poor Law in 1834 was, by common consent, 'concerned as much with government as it was with poverty'.[1] This problem of government encompassed questions of moral regulation as well as administrative efficiency. The architects of the 1834 reform wanted to transform the moral economy of pauperism; Edwin Chadwick, for example, described the new Poor Law as an administrative 'experiment' in the treatment of a 'moral plague'.[2] Terms such as these indicate that Poor Law reform needs to be seen in a broad political and intellectual context; that context is provided here by a discussion of the changing nature of mid-Victorian government. For some historians, of course, the middle decades of the nineteenth century mark the high point of the ideology of *laissez-faire*, which is supposed to have inspired a reduction in the powers of the central state. The argument in this chapter suggests that this view is a partial one, not least because the advance of 'free market' ideology on some fronts was accompanied by the intensification of state regulation on others.

It is sometimes argued that the reformers of 1834 effectively confirmed the right of paupers to relief, by rejecting calls for the entire abolition of the Poor Laws. A better way of putting this, as one official subsequently acknowledged, is to say that the 1834 law affirmed the duty of the state to relieve the poor under certain conditions.[3] The sociologist Georg Simmel once described this fine distinction between right to relief and duty to relieve as fundamental to the modern conception of public assistance; in the case of the new Poor Law, for example, 'all the relations between obligations and rights are located, so to speak, above and beyond the poor'.[4] The 1834 reform underscored the rights and duties of property, not those of individual paupers. In fact, the Poor Law reformers themselves had little time for the idea of natural rights and natural justice. Bentham himself had described the idea of moral rights as 'nonsense on stilts', a phrase for which history has still not forgiven him.[5] The relief of the poor was seen as a matter of expediency, of governance, and not of right. The function of a reformed Poor Law was to encourage self-exertion by forcing 'free' labour onto a competitive market. This required, paradoxically, a dramatic extension of the powers of the central state. On the one hand, then, the reformers constructed a world of autonomous individuals contracting freely in an open labour market; on the other, they envisioned a landscape of moral discipline and government. The 1834 reform illustrates perfectly the janus-face of modern liberalism.[6]

## *Laissez-faire* and the liberal state

In general accounts of the changing functions of the British state, the mid-nineteenth century has frequently been treated as an era of *laissez-faire*. Typically, measures such as the new Poor Law or Corn Law repeal are portrayed as legislative expressions of classical liberal principles, subordinating the activity of the state to the needs of self-regulating markets. The mid-Victorian state is thus characterised as a 'nightwatchman', operating on the sidelines of a largely unregulated economy.[7] (The modern state, in contrast, is said to have intervened so extensively in economic and social affairs that it no longer seems appropriate to talk of 'markets' in isolation from their institutional supports).[8] This view of mid-Victorian government echoes the claims of nineteenth-century free-traders, who were not slow to contrast the British preference for 'non-interventionism' with the statism of continental Europe. Historians have frequently resorted to similar comparisons in order to sustain particular views of the 'peculiarity' of the British state; indeed, it is sometimes claimed that the lack of state interventionism was a critical factor in the unrivalled expansion of the domestic economy. Recent studies suggest, however, that it was not absolute weakness but relative efficiency which characterised the British state during the seventeenth and eighteenth centuries.[9] And far from being denuded of its powers during the heyday of *laissez-faire*, central government was to augment its infrastructural capacities in rather significant ways.

As an ideology, *laissez-faire* depends on a powerful myth – the 'free market', represented as an autonomous domain, in which production and exchange take place with a minimum of interference from agencies other than buyers and sellers of commodities. This image of the market as a place of free exchange between individuals is perhaps the most significant legacy of nineteenth-century liberalism; indeed, the market metaphor has survived the assault of classical political economy's most influential critics. Even Marx, who claimed to have de-naturalised the market by exposing its historical and social conditions of existence, frequently portrayed its workings under capitalism as in some senses spontaneous and autonomous. Such a view necessarily characterises state activity as 'intervention', disturbing or rectifying essentially economic processes. In doing so, it may obscure the extent to which the state actually constructs and protects the 'market', both institutionally and ideologically. As Corrigan and Sayer observe in their study of English state formation, 'State intervention (though the term is almost unusable if we are to understand what took place) enabled, accomplished, stabilized, regulated into dominance the market on which *laissez-faire* theory depends'.[10] The case of the 1834 Poor Law reform provides probably the best example of the process by which a 'free market' (in human labour) was policed by a newly-formed apparatus of state. It is for this reason that the reform provides the centrepiece of Karl Polanyi's magisterial account of the commodification of labour in *The great transformation*. For Polanyi, 1834 demonstrates the extent to which the labour market is (literally) a political fiction.[11]

The web of ideas associated with *laissez-faire* – emphasising the moral virtues of individual self-exertion, the sanctity of private property and the vices of political centralisation – clearly helped to shape the ideological framework in which the business of mid-Victorian government took place. The manner of this influence remains a subject of considerable debate. The controversies over individualism and collectivism which shook the British intelligentsia at the end of the nineteenth century have cast a long shadow over historical writing on the role of the state in the preceding period. In the course of these constitutional and philosophical debates, liberals, conservatives and Fabians alike tended to portray mid-Victorian Britain as the *locus classicus* of both 'individualism' and *laissez-faire*. A. V. Dicey, the author of an influential study of *Law and opinion* (1905)[12], marked out the period between 1820 and 1870 as one of individualism, during which the power of the central state was severely restricted. Although their political conclusions were markedly different, Fabians such as Beatrice and Sidney Webb similarly assumed that individualism placed important limits on state power during the mid-nineteenth century. One of the legacies of these influential interpretations was, as Harold Perkin has shown[13], the common identification of the philosophy of individualism with the policy of *laissez-faire*, and that of collectivism with state intervention. Perkin argues that such equations are misleading, for even the purest philosophies of 'individualism' may demand considerable levels of state regulation in order to secure the freedom of individuals and the defence of private property. This argument has important implications for our understanding of reforms such as the new Poor Law.

In Dicey's work, the paradigm case of mid-Victorian individualism was assumed to be the Benthamite philosophy of utilitarianism. Bentham's principle of 'universal egoism' or 'self-preference' asserted that the root motive for all human action was individual self-interest. To this extent Bentham's moral philosophy may certainly be described as 'individualist'. However, as Halévy points out, the ultimate aim of Bentham's political philosophy was to place 'each of the members of the political society in social conditions such that his own private interest shall coincide with the general interest'.[14] In a society of individuals, Bentham argued, there was no natural or spontaneous identity of interests; therefore the greatest happiness of the greatest number (the popular maxim of utilitarianism) could only be achieved through the construction of methodical and efficient laws governing human affairs. There was thus a powerful rationale within utilitarianism for government action; while Bentham's moral philosophy was grounded in individualism, his political philosophy led him towards increasingly regulative solutions to social problems. His blueprint for a utilitarian state, compiled towards the end of his life in the *Constitutional code*, was designed to implement principles of rational administration and policy. It contemplated the establishment of a number of new central government departments, responsible for such things as Interior Communications, Indigence Relief, Education and Health.[15] The Benthamite state thus resembles a machine in which the individual parts are assembled in a way which secures the interests of the community at large rather than those of particular individuals. As Halévy writes,

> The state, as conceived by Bentham, is a machine so well constructed that every individual, taken individually, cannot for one instant escape from the control of all the individuals collectively.[16]

This vision of the Benthamite state is clearly at odds with Dicey's. It suggests that, at least in Bentham's case, individualism as a philosophy certainly did not require *laissez-faire* as a policy. It might be argued that Bentham was a special case; indeed, several of Dicey's critics point out that Bentham's 'individualism' was of a very special kind.[17] Yet there is sufficient evidence to suggest that many other contemporary philosophers of individualism were far from hostile to regulation in general.[18] In the writings of the classical political economists, for example, government was frequently accorded an important role in the construction of legal, fiscal and moral frameworks which were to regulate individual conduct and protect private property. Furthermore, even the most ardent champions of *laissez-faire* in the sphere of international trade accepted the need for state intervention in other areas of policy, especially where social problems were concerned. The political economist J. R. McCulloch once observed that blind faith in *laissez-faire* savoured 'more of the policy of a parrot than a statesman or philosopher'.[19] Nassau Senior, one of the architects of the 1834 Poor Law reform, embraced a similar view: 'The only foundation of government is expediency – the general benefit of a community. It is the duty of government to do whatever is conducive to the welfare of the governed'.[20]

The concept of *laissez-faire*, taken at face value, clearly masks the increasing scale of state regulation during the nineteenth century. In recent years, historians have paid increasing attention to the nature of government growth during this period. What has sometimes been neglected in this literature is the extent to which state regulation complemented (rather than contradicted) the ideology of the free market.[21] As Raymond Grew has pointed out, the state had an increasingly important role in policing the boundaries between different 'fields of play' in the economies and societies of nineteenth-century Europe.[22] In Britain, for example, official regulation was frequently deemed necessary to maintain and protect the free space of the market by encouraging 'independent labour', restricting the activities of trades unions, removing protective tariffs and discouraging monopolies. Such policies often required an expansion of the bureaucratic apparatus of the state, just when it was being claimed that government was pursuing a policy of 'non-intervention'. As Polanyi observes,

> The road to the free market was opened and kept open by an enormous increase in continuous, centrally-organised and controlled interventionism.[23]

Nowhere was this more clear than in the case of the 1834 Poor Law reform. The next section provides a brief account of the reform and its context. There follows a more general discussion of the process of 'centrally-organised and controlled interventionism' which Polanyi associates, paradoxically, with the triumph of the free market.

## The 1834 Poor Law reform

The Report of the Royal Commission which preceded the 1834 Poor Law has rightly been described as 'one of the classic documents of Western social history'.[24] For the reformers, it was a beacon of enlightenment, pointing unambiguously towards rational principles of administration. For their critics, the document served only to legitimise an experiment in studied brutality. Amongst historians, judgements have been similarly varied. While some have portrayed the Report as a neo-Malthusian assault on government intervention in the labour market, others have represented it as laying the foundations for the welfare state. In short, the 1834 Report has been interpreted and explained in a thousand different ways ever since it was hurriedly compiled in the Spring of 1834. Such a diversity of readings perhaps reflects the fact that the reformers of 1834 raised, but failed to resolve, fundamental questions about social order and moral authority in an era of liberalism.

The appointment of a Royal Commission of Inquiry in 1832 was by no means the first attempt to reform the Poor Law. Parliamentary committees had addressed the problem of pauperism on many occasions since the end of the French wars, and some legislation of a permissive character had followed. At local level, furthermore, a host of small-scale experiments in relief policy had already anticipated many of the measures adopted in 1834. Yet, as one historian remarks, reform before 1834 had a 'stuttering quality';[25] it was instituted and abandoned according to short-term and local requirements. The old Poor Law rested on a decentralised system of government operating under the umbrella of a complex body of legislation which (according to one of Chadwick's biographers) 'pleased few and was understood by nobody'.[26] Although innovation under the old system was possible, the reformers of 1834 argued that it was inevitably compromised by the absence of any 'fixed and general plan'.[27] Indeed, the experimental adoption of local measures could create new obstacles to radical reform, as in the case of parishes incorporated under Local Acts (to be discussed in chapter 3). What was advocated in 1834 was, in contrast, national reform based on universal principles.

The Poor Law Commission was established in February 1832. Following the appointment of the seven (later nine) members of the Commission, twenty-six Assistant Commissioners were dispatched to various parts of England and Wales in order to gather local information. They were issued with a set of specific instructions which encouraged them to focus their investigations on specific abuses and particular remedies. In addition to these personal visits, two sets of written queries were sent out to rural and urban parishes (about one in ten of which were returned completed). The Commissioners published their Report in February 1834, and followed it several months later with a voluminous set of appendices. As this document was to acquire talisman status with the Poor Law authorities after 1834, it is worth paying close attention to the diagnosis it offered, and the solution it proposed.

Two of the Commissioners appear to have had a disproportionate influence on the drafting of the Report; Nassau Senior, Professor of Political Economy at Oxford, and Edwin Chadwick, protêgé of Jeremy Bentham. Both assumed, like Bentham, that the purpose of a Poor Law was not the abolition of poverty, since poverty was the natural and proper condition of all those who had to work to stay alive; as Bentham put it, 'as labour is the source of all wealth, so poverty is of labour. *Banish poverty, you banish wealth*'.[28] The object of a Poor Law was rather the relief (and prevention) of indigence – i.e. that condition whereby individuals were unable to secure their survival in the labour market. Bentham's arguments on this point were copied almost word for word into the relevant sections of the 1834 Report, and were subsequently recycled in Chadwick's anonymous defence of the new law, published in 1836.[29] The goal of the reformers was to police and protect the labour market, not to replace it. They contended that local administrators under the old Poor Law had effectively undermined the 'independence' of free labour. The Report thus identified 'the great source of abuse' as the granting of relief to the able-bodied male labourer already in receipt of wages, through weekly supplements (the 'Speenhamland system') or through the mechanism of labour-rates.[30] This had proved fatal to both the self-discipline of labourers and the operation of a free labour market.

The diagnosis presented in the Report was to a large extent anticipated by Nassau Senior four years earlier. In a discursive preface to his *Three lectures on the rate of wages* (1830), Senior had attributed popular disturbances in the south of England to the effects of the allowance and roundsmen systems of relief, which he claimed were disrupting the relations between employer and employee. The distinction between the slave and the free labourer drawn to illustrate this argument was also to appear in the 1834 Report. The slave, entitled to maintenance in return for his labour, was 'equally incapable of being benefited by self-restraint or injured by improvidence'; the freeman, in contrast, was 'the master of his exertions, and of his residence'. The Poor Laws had begun to confuse the two:

> The labourer is to be a free agent, but without the hazards of free agency; to be free from coercion, but to enjoy the assured subsistence of the slave. He is expected to be diligent, though he has no fear of want; provident, though his pay rises as his family increases; attached to a master who employs him in pursuance of a vestry resolution; and grateful for the allowance which the magistrates order him as a right.[31]

Senior attributed a variety of moral evils to the new systems of relief, arguing that 'the instant the labourer is paid not according to his *value*, but his *wants* he ceases to be a free man'.[32] The alternative to the discourse of rights associated with the old Poor Law was provided by the language of contract. The moral imperatives of free market liberalism demanded self-exertion and self-restraint. A reformed Poor Law ought to force labour onto a 'free' market.

This argument was copied, almost word for word, into the 1834 Report. The allowance system and relief-in-aid of wages were said actively to promote pauperism, punishing all attempts at 'self-reliance', giving the labourer 'all the slave's security for subsistence without his liability for punishment'.[33] The alternative solution was to remove any incentive to pauperism. 'The first and most essential of all conditions', the Report famously exclaimed, 'is that [the pauper's] situation on the whole shall not be made really or apparently so eligible as the situation of the independent labourer of the lowest class'.[34] The entire strategy of 1834 thus rested on the drawing of 'a broad line of distinction between the class of independent labourers and the class of paupers'.[35] The best way to police this line, according to the reformers, was by making the provision of relief to the able-bodied male labourer dependent on entry into 'a well-regulated work-house'. This would not only provide a clear distinction between 'independent' and 'dependent' labour, as far as the able-bodied were concerned, but would also constitute a 'self-acting test of the claim of the applicant'. The prospect of a managed and disciplined existence in one of these institutions would, it was contended, deter all but the truly destitute from applying for relief.

The publication of the Commissioners' Report was followed by the introduction of a Poor Law Bill into the House of Commons. The Bill, which became law in August 1834, was (in contrast to the Report) largely concerned with the administrative mechanisms required to establish a new machinery of Poor Law government. It enabled the appointment of a new central Poor Law authority, with powers to amalgamate local parishes into larger administrative units (Poor Law Unions) and to issue rules regulating relief policy and practice. It also allowed for the appointment of permanent inspectors, at first called Assistant Commissioners, to supervise the implementation of the new system in the localities (see chapter 3). The speedy passage of this legislation through Parliament suggests that, initially, the reform met with little opposition. Yet there were those, even in 1834, who regarded the Commission's Report as a backward-looking attempt to reinforce existing patterns of privilege and wealth. The London radical John Wade, for example, condemned the Report for its 'singular want of comprehensiveness of view – a disregard of general principles – and an absence of correct information on the character and condition of the labouring classes'.[36] Wade complained that the Commissioners had virtually ignored the plight of the poor in the industrial towns and cities, concentrating instead upon a small number of 'abuses' in agricultural districts. This charge of selectivity has been echoed by economic historians such as Mark Blaug, who have re-examined the statistical evidence available in the appendices to the 1834 Report and suggested a variety of alternative explanations which 'fit' the data.[37] To these historians, the economic irrationality of the old Poor Law (as measured by the effects of out-door relief on wage levels and productivity) was far from conclusively demonstrated in 1834.

Despite the impression given by its voluminous appendices, the 1834 Report did not offer a detailed statistical justification of its conclusions. It is anachronistic to treat the Report as an early nineteenth-century version of a twentieth-century model of inquiry, not least because (as Karel Williams has argued) its use of data was illustrative rather than inferential.[38] The reformers were concerned above all with the moral consequences of existing relief policies, consequences which (according to Nassau Senior) 'cannot be expressed in figures or tables'.[39] The key to the reformers' approach therefore lies less in their analysis of quantitative data than in their *a priori* moral calculus concerning the pleasures that should attach to independent labour and the pains that should correspond to dependent pauperism. The ultimate aim of Poor Law reform, as of so many nineteenth-century projects of moral regulation, was to inspire self-discipline and to promote the moral authority of government; at the centre of this strategy was the figure of the 'independent labourer'.

Poor Law reform in 1834 should be seen in the wider context of a crisis of social authority reverberating throughout rural England during the early 1830s; as Peter Dunkley has pointed out, 'vital questions of discipline and order' were at stake.[40] The unprecedented upheavals of these years effectively turned the long debate over Poor Law reform into a question of considerable urgency. 'Lurking behind the financial concerns of peers and squires', according to another historian, 'was the spectre of social disintegration'.[41] It was this sense of crisis which propelled a revolutionary administrative reform through Parliament. Dunkley suggests that 1834 marks a 'watershed in the history of authority'.[42] Such a judgement is, in the present context, a little premature. For the consequences of the new Poor Law could not have been entirely foreseen in 1834. The reformers designed a system; but they could not dictate precisely how it was to be applied. The final section of this chapter considers the development of new techniques of government during the middle decades of the nineteenth century.

**Revolutions in government?**

The changing character of government regulation during the mid-nineteenth century has been a subject of considerable debate amongst historians. Until relatively recently, the historical literature on nineteenth-century social administration has been dominated by reappraisals of the work of A. V. Dicey. As we have seen (above), some historians dispute Dicey's interpretation of Benthamism as a philosophy of 'individualism'; others have rejected the more general idealist assumptions (concerning the relationship between 'law' and 'opinion') which underlay his work.[43] The most influential alternative account of changing patterns of central regulation is provided in Macdonagh's model of the 'revolution in government', first published in 1958, which characterised the growth of government in terms of a developmental sequence of stages.[44] The roots of government growth between 1825 and 1875, he argued, typically lay in the sensational 'exposure' of some intolerable social problem or scandal. This was often followed by the passing of permissive legislation, enabling but not dictating a limited degree of local reform. The inadequacies of these initial attempts at reform frequently provided a rationale for the creation of new and permanent central authorities with their own mechanisms of supervision, in turn laying the basis for further compulsory legislation and routine intervention. Gradually, according to Macdonagh, the centralisation of control and expertise would ensure the development of a new style of state regulation.

Macdonagh's account, based initially on research into the administration of the Passenger Acts, was intended to serve as a generalised model of government growth. It inspired a succession of historians to undertake detailed research on administrative change within particular departments of state, including the Education Department, the Board of Trade and the Local Government Board.[45] Although it remains an important landmark in the history of social administration, Macdonagh's model has attracted a great deal of criticism since its initial formulation. Of all its limitations, there are two which are of particular relevance in the present context. The first derives from its emphasis on pragmatism rather than ideology as the root cause of government growth. In stark contrast to Dicey, Macdonagh found little place for the systematic impact of philosophies such as utilitarianism on the practice of government. However, there is enough contrary evidence to suppose that all sorts of philosophies and theories (utilitarianism, political economy, associationism, medical theory, and so on) did have important impacts on the form and direction of central policies, though not necessarily in the instrumental manner that Macdonagh envisaged.[46] The challenge here is to show the various ways in which discourses about social policy framed specific programmes and practices of government. Secondly, Macdonagh's central government departments are depicted as strangely olympian institutions, divorced from the day-to-day politics which surrounded them. It is the unanswerable pressure of 'social evils' which provides their cue; and their subsequent evolution is interpreted as an almost automatic process, driven by the cycle of inquiry, reform and inspection. As several critics have pointed out, such a view comes dangerously close to confirming the assumptions of the most inward-looking administrative histories.[47] In contrast, the present account pays particular attention to the ideological and political significance of changing modes of central regulation.

The period between 1830 and 1860 marks an important moment in the history of government practice; one which might be seen in terms of a marked extension of what Mann calls the 'infrastructural' capacities of the central state (see chapter 1). The Poor Law Commission appointed in 1832 provided a model for many subsequent official inquiries in the fields of employment, local government, health, policing and welfare. Macdonagh's critics point to the influence of a relatively small group of social reformers in shaping the form and outcome of these inquiries. Henriques, for example, identifies a distinctively Benthamite 'machinery' of inquiry which she claims was common to a wide variety of fields.[48] She emphasises the increasing use of Royal Commissions as a strategy of reform; during the 1850s, for example, no less than seventy-four new Commissions were appointed.[49] In some ways, the Royal Commission provided an ideal vehicle for the propagandists of social reform. In contrast to the Select Committee, the traditional mechanism of Parliamentary inquiry, the membership of Royal Commissions was not confined to the Commons and the Lords. Another characteristic feature of government inquiry during the 1830s and 1840s was the appointment of teams of Assistant Commissioners to conduct local investigations, a procedure which vastly expanded the information-gathering capacities of reformers. The result, Henriques argues, was less objective policy-making than effective propaganda. With a little influence in the right places, the membership of Commissions could be packed, witnesses could be pre-selected, evidence could be manipulated and reports selectively disseminated.[50]

Social reformers like Edwin Chadwick called not simply for more enlightened laws; they sought a fundamental reorganisation in the machinery of government. Perhaps the most significant feature of the 'revolution in government' during this period lay in the appointment of new permanent central authorities. On several occasions, the Report of a Royal Commission of Inquiry led directly to the establishment of entirely new agencies of state. In some cases, the new authorities were attached to existing departments, as in the case of the Factory Inspectorate appointed in 1833 (which was incorporated within the Home Office); in others, they were set up as independent entities, only indirectly answerable to Parliament, as in the case of the Poor Law Commission and the General Board of Health. The rationale for the establishment of new central authorities such as these was, broadly speaking, administrative. The 1834 Poor Law Report, for example, clearly envisaged the formation of an integrated *system* of Poor Law government in place of the tangled web of parochial authorities which existed under the old law (see chapter 3). Although the Poor Law Commission established in 1834 was initially authorised to function for five years, its lifespan was subsequently extended. The new authority was to stress its independence from 'local interests'; its commanding view over the whole country, reformers suggested, would enable a much greater degree of uniformity in administrative practice. As the philosopher John Austin declared in 1847, 'Uniformity simplifies the administrative machine, lays it bare to discerning observation and criticism; and this enables the central authority to watch and control its movements with comparative facility and effect'.[51] This argument reflected the gathering momentum behind calls for the 'centralisation of knowledge';[52] in the case of Poor Law reform, as we shall see in chapter 3, this was to take the form of a new and permanent state apparatus of information-gathering and inspection. This was a classic case of surveillance being improved in order to enable the routine monitoring of administrative practice (see chapter 1).

There was something revolutionary in the establishment of centrally-based inspectorates during the mid-nineteenth century. The principle of inspection itself was not new; the supervision of Justices of the Peace had long constituted a decentralised form of surveillance, watching over local government and society'.[53] What was unprecedented was the specific form of the new inspectorates: central *corps* of salaried specialists, appointed to gather local intelligence and enforce official standards. One administrative scientist describes the new officials as the 'in-between' men and women of state, key links between central and local authorities.[54] Between 1832 and 1875, over twenty separate central inspectorates were established, in fields ranging from workhouse education to salmon fisheries. Although their specific duties varied considerably, these inspectorates shared a number of general functions. All were appointed to supervise or enforce the local administration of particular laws or official standards; all were expected to make periodic reports to the centre on the state of local administration; and all claimed some kind of 'expertise' in their particular field. These 'statesmen in disguise'[55] were to have an important, some would argue pre-eminent, influence on the direction of subsequent social policy.

The introduction of routine official inspection was designed to ensure the coordination of local practice with central policy. Many of the new inspectors, including those appointed by the Poor Law Commission, were responsible for a particular region; these were the field officers, the prime means of personal communication between central and local authorities. In addition, the inspectorates served as 'transmission belts'[56] for the diffusion of particular models and policies, perpetually challenging local administrators to expand their horizons. Central inspection was thus intended to set in motion a self-perpetuating process of inquiry and reform. This, at least, was the theory. In practice, however, the role of individual inspectors was frequently compromised by the lack of both adequate financial support and an effective machinery of enforcement.[57] The extent to which this was true varied between different spheres of social policy. The Poor Law inspectorate, for example, was to be particularly active during the early years of implementation in the 1830s and early 1840s, when the Assistant Commissioners were almost constantly on the move (see Figure 2.1). Subsequently, however, individual inspectors were to take a far less prominent role in the development of policy. The controversies surrounding the Commission during the 1840s made officials increasingly wary of adverse publicity (see chapter 3). Furthermore, the establishment of new centres of 'expertise' (in other departments of state) effectively eroded the autonomy of the Poor Law inspectors themselves. In the case of the Poor Law, therefore, what Roy MacLeod calls the 'heroic' phase of central inspection[58] was to be relatively short-lived. It would be wrong to treat this as a simple retreat from central regulation. While the charismatic role of individual inspectors was to decline, the routine regulation of local administration was to proceed unabated. The burgeoning bureaucracy of central regulation reflected a more general process (presenting what MacLeod calls a 'dilemma of departmentalisation') by which the expertise of specialists was increasingly subjected to the disciplines of legal and administrative control.[59]

Compared with the surveillance capacities of modern governments, the effectiveness of nineteenth-century central inspection was certainly limited. Yet to suggest that inspection was relatively inefficient is not to say that it

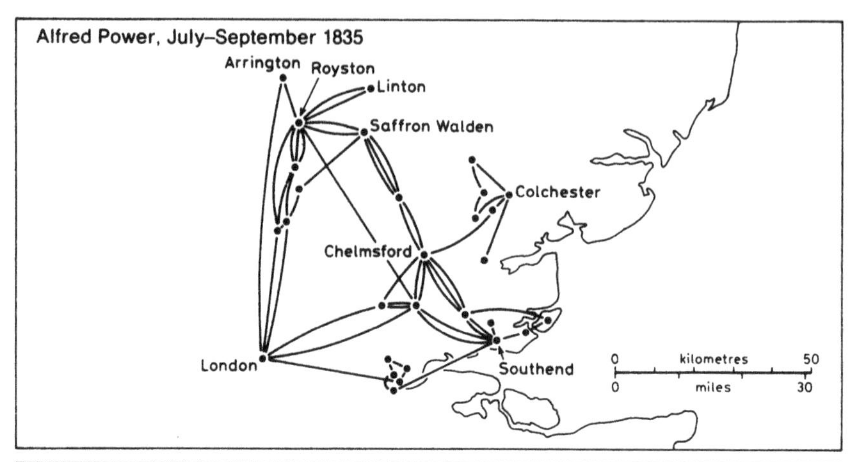


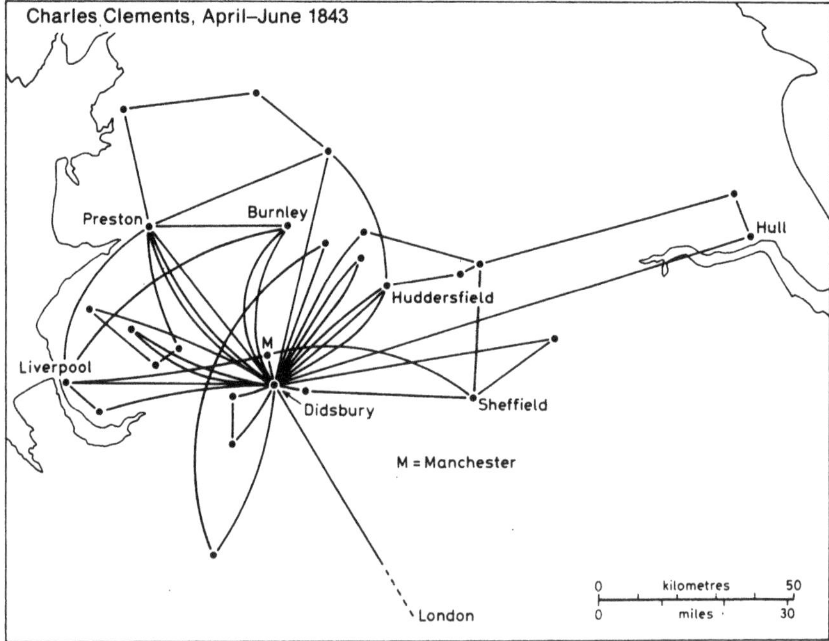


Figure 2.1 Circuits of two Assistant Poor Law Commissioners, 1835 and 1843 (*Sources*: MH 32/63; Parliamentary Return, PP 1845 XXXVIII)

was insignificant, in either instrumental or symbolic terms.[60] As noted above, the literature on the 'revolution in government' has frequently failed to acknowledge the ideological and political ramifications of central government inspection during the mid-nineteenth century. As one critic has argued, the mid-Victorian state has sometimes been presented as though it was 'an autonomous and benign agency extending its services to meet social needs'.[61] Yet both the definition of these 'needs' and the means of their fulfilment raised far-reaching questions of ideology and politics. While many of the inspectors portrayed themselves as standard-bearers of rational reform against local obduracy and vested interests, they were far from 'disinterested professionals'.[62] Contemporary controversies over centralisation propelled the new inspectorates to the centre stage of popular politics during the 1830s and 1840s. In the case of the Poor Law inspectorate, at least, the sensitivity of their task was not in doubt. As the Commissioners advised their newly-appointed inspectors in 1834,

> The peculiar nature of your office gives you opportunities belonging to no other public functionaries, to acquaint yourself with the condition of the working class.[63]

The encounter between these representatives of the state and the working class was not altogether a happy one, as we shall see (in chapter 7). Nevertheless, the ideological turbulence which marked the early years of central inspection was subsequently to subside, giving way to quieter times, when the work of inspectors was less likely to arouse deep-seated political objections. This might perhaps be seen as a reflection of the changed political mood of the 1850s and 1860s, the heyday of local self-government,[64] when ardent centralisers like Edwin Chadwick were losing their influence in Whitehall. Yet there were others, such as John Simon at the Privy Council,[65] who were proceeding in quieter, though no less effective ways. The 'anti-centralisation' of the middle decades of the nineteenth century might thus be profitably compared with the 'anti-imperialism' of the same era; though grand designs were frequently disavowed, the practical drift of policy was in the direction of more intervention rather than less. The later decades of the nineteenth century witnessed a renewed phase of government growth, marked by the adoption of routine bureaucratic practices in ever-expanding departments like the Home Office and the Local Government Board.[66] By the end of the century, central inspection – the infrastructural power of the state – had become a part of the fabric of everyday life. 'Throughout the journey of life', mused one Local Government Board inspector in 1909, the citizen would find 'an Inspector perched on every milestone'. As it was in life, so too in death; 'for is there not the Inspector of Burial Grounds?'[67]

# 3

# Rational landscapes? The geography of Poor Law government

> When this change occurred, whereby assistance to the poor became an abstract obligation of the state – in England in 1834, in Germany since the middle of the nineteenth century – . . . local organisation [was] made into a mere technique in order to attain the best result possible; the municipality is no longer the point of departure, but rather a point of transmission in the process of assistance.
>
> Georg Simmel (1908)[1]

It is as well to preface a survey of the administrative landscape of the new Poor Law with a reminder of the violent political events which marked its formation. Administrative reforms, even in peaceful times, require sensitive adjustments in the status and influence of various individuals and groups; these power relations cannot simply be wished away. During the 1830s, administrative questions were intensely political. They provided a focus for those representatives of the middle classes who wanted a greater voice in the political affairs of the nation. And they were central to wider debates over the government of the working classes. Nowhere is this clearer than in the case of Poor Law reform. In the words of one historian,

> The Poor Law filled the whole horizon in 1834. And here, there, and everywhere, were Chadwick's young crusaders, the Assistant Commissioners, scouring the country in stage-coaches or post-chaises, or beating up against the storm on ponies in the Weald, returning to London, their wallets stuffed with the Tabular data so dear to Philosophic Radicals, to draft their sovereign's decrees declaring the union and stating his austere principles of administration, and then back to see that they were carried out.[2]

The impact of these events can be assessed in several different ways. While economic historians have focussed on the economic conditions and consequences of the 1834 reform, social historians have debated the extent to which the new Poor Law reflected broader changes in the balance of social power. Several of the Poor law reformers themselves described their proposals in terms of radical social change; Nassau Senior, for example, portrayed the new law as an attempt to 'dethrone' the landed interest.[3] Critics were less sanguine, suspecting that beneath the reformist rhetoric the rights of property and privilege had been strengthened, not assailed.[4] It is now widely assumed that the 1834 reform ought not to be seen (as Marx once described it) as an assault on landed power by the newly-enfranchised industrial bourgeoisie. The reform was devised primarily for the agricultural districts of England and Wales, and was supported as much by the influential sections of the landed gentry as by the philosophical radicals. The 1834 Act therefore symbolised not the triumph of 'market' values *against* the landed gentry, but the final stage of the incorporation of these values into the governing strategy of a broadening elite.[5] Anthony Brundage has gone further to claim that, far from 'dethroning' the landed interest, the new Poor Law provided the gentry with new power bases in a property-centred system of local administration.[6] The extent to which this was so depended on the social geography of local influence; the character of Boards of Guardians in Northamptonshire, for example, was quite different to those of London or Wales. It is also important to recognise that the new system redefined the boundaries of 'local' authority; individual Guardians were simply unable to exercise the kind of influence at the disposal of landowning magistrates under the old Poor Law.

In assessing the significance of 1834 for subsequent policy, it is tempting but ultimately misleading to focus on the arguments of the reformers alone.[7] The most obvious flaw in such a strategy is that the rationale for reform presented in the 1834 Report and its subsequent interpretation are not necessarily identical. The Report provided a critique of existing practice and some guidelines for its reform; yet the new Poor Law itself was to be administered in a world of conflict and negotiation, strategy and counter-strategy. Interpretations of 1834 were to be radically different in 1837, and again in 1867; new contexts brought forth new diagnoses.[8] For this reason alone, the history of Poor Law administration after 1834 cannot be seen as an expression of the essential message embedded in the Poor Law Report. That document gave authority to a new language of social administration; but it did not pre-determine the shape of the new Poor Law system over the next fifty years. In order to understand the evolution of this new system, it is necessary to consider the evolving geography of Poor Law government after 1834.

### The machinery of Poor Law government

The reformers of 1834 anticipated a revolution in Poor Law government. At the centre, there was to be a new central authority, the Poor Law Commission, armed with extensive powers of inspection and intervention; at local level, an entirely new tier of government, transcending the petty interests of the parish. This, at least, was the theory; and the reformers knew that efficient communication between the various parts of this system was the key to its success. Theirs was not simply a call for more centralisation; it was an attempt to alter the very framework within which decisions were made. This is what Chadwick meant when, almost in passing, he described the 1834 Act as 'an administrative law';[9] for the first time, poor relief and paupers were to be organised within an *administrative system*.

After its establishment in 1834, the new central Poor Law authority embarked on the task of constructing this new system. Orders were issued governing the conduct of Guardians' elections, the duties of local officials, the management of Union workhouses and the granting of relief, and hundreds of standard forms and accounts were issued to the new Guardians, Relieving Officers, Medical Officers, Assistant Overseers and workhouse officers.[10] (The formation of Poor Law Unions is considered separately below.) The consolidation of these Orders in 1847 did not halt the flow of new regulations, which were subsequently to prescribe, amongst other things, the proper amount of tobacco to be given to each workhouse inmate and the temperature at which invalids were to be bathed. The significance of these regulations for our present purpose lies in their form rather than in their content. The aim was to make the process of granting (or withholding) relief an administrative question, determined by fixed rules, rather than a personal one, influenced by 'partial' and capricious motives. As Martin Wiener has recently pointed out, this quest for uniformity was an important impulse in many other spheres of social reform besides the new Poor Law.[11]

The issuing of formal rules was only one of the ways in which the power of the centre was to be felt in the localities. In 1834, the new Poor Law Commission was given the power to make routine inspections of local administration. The Commissioners described their representatives, the Assistant Commissioners (known officially as inspectors after 1847), as their 'eyes, hands and voice'; without their regular visits to individual Unions, they noted, the centre would be left with 'duties of control and regulation without the means of observation or the power of investigation'.[12] Administrative surveillance required the routine monitoring of the activities of local authorities and local officials (see chapter 2). The functions of the Assistant Commissioners were thus to include regular attendances at Guardians' meetings, routine correspondence with local authorities, the organisation of special inquiries into cases of neglect or abuse, the inspection of workhouses and the writing of periodic reports on pauperism and Poor Law administration.[13]

Central inspection was the lynchpin of the post-1834 system. In the early years, much depended on the diplomatic skills of the Assistant Commissioners, who often faced considerable local hostility. While many of them portrayed themselves as ambassadors of rationality in the battle against parochial obduracy and vested interests, it would be wrong to accept such claims at face value, especially given the highly charged political atmosphere of the 1830s. Their role during this and subsequent periods was one of influence as much as enlightenment. Amongst those appointed in 1834, lawyers, magistrates and military men predominated;[14] figures of authority as much as expertise. Their initial task was to set the law in motion by forming Poor Law Unions and establishing Boards of Guardians. Thereafter, as far as central government was concerned, what was wanted was routine control, requiring fewer local visits; the number of district inspectors, which had been above twenty at one time, was reduced to twelve in 1841 and, by 1847, there were only nine left. This diminution in the size of the inspectorate increased the workload of individual inspectors; a comparison of the movements of two inspectors in Figure 2.1 suggests a transition to a less intensive form of inspection following the initial stages of implementation. During the 1840s, Poor Law inspectors were almost permanently on the move, with only occasional periods of leave or attendance at the central office.

In addition to the hostility of anti-Poor Law campaigners and the obduracy of local Guardians, the first generation of Poor Law inspectors was faced with increasing divisions within the central authority itself. As the atmosphere at Somerset House became increasingly acrimonious, one wit described the relationship between the senior Commissioner and his Secretary (Edwin Chadwick) as 'an ill-assorted marriage of a Catholic husband and a Protestant wife, the wife somewhat the cleverer of the two, but with no privilege beyond the use of her tongue'.[15] Several of the Assistant Commissioners shared Chadwick's doubts about the resolve of the Commissioners. Charles Mott and William Day, whose resignations were forced in 1842 and 1843, following disputes with Boards of Guardians in the West Riding and Wales, both complained of a lack of support from their superiors.[16] But their complaints were as nothing compared to the torrent of abuse heaped on the Commissioners by one of their colleagues, H. W. Parker, whose resignation was forced during the infamous Andover scandal in 1845. The Select Committee inquiry into the treatment of paupers at Andover became, in effect, a trial of the central authorities rather than of the Guardians at Andover. With its embarrassing revelations of disunity within the Commission itself, this episode provided further evidence for those calling for the reorganisation of the central authority.

Following the creation of a new Poor Law Board in 1847, the process of inspection became more routine. During the 1850s and 1860s, debates over the quality of workhouse provision raised questions about the competence of Poor Law inspectors in matters of education and health, and relations with officials in other government departments were frequently strained. The Poor Law as a whole was increasingly cast as a bureaucratic regime, unable to adapt to the special needs of specific groups such as the sick or the young.

To some extent, the transfer of the Poor Law schools inspectorate from the Education Department to the Poor Law Board, and the creation – in 1871 – of the Local Government Board (in which the Poor Law officials soon gained the upper hand)[17] disabled such embarrassing assaults on the inspectorate's expertise. After 1871, the administrative style of the Local Government Board provided a new context for Poor Law inspection. The inspector's activities were far less in the public eye than those of his predecessors; his or (very rarely) her authority derived less from social standing than from administrative competence and an ability to master the vast mass of regulations accumulated since 1834; and the expertise of an increasing number of specialist colleagues was readily available. During the 1830s, Poor Law inspectors had been highly visible, sometimes charismatic, figures in a landscape of political conflict; by the 1880s, their role had become a less prominent one, confined by an increasingly legal-bureaucratic administrative culture.[18] The violence of the early phase of implementation had given way to what one inspector described as a 'paper war'.[19] In this respect, the history of central inspection under the new Poor Law seems to mirror the general pattern outlined in chapter 2.

One of the most significant powers available to the new central authority after 1834 was a financial one. The 1834 Act gave the Commissioners the power to withhold their sanction from any local expenditure which they considered illegal or contrary to an official Order. Union accounts were to be independently inspected by auditors answerable to the central authority. At first these officers were elected by individual Boards of Guardians, but after 1844 they were appointed on a district basis; in 1868, the central authority gained the right of appointment itself. The district auditors were given powers of disallowance and surcharge on individual officers who had spent money in a way they regarded as unlawful or contrary to an Order. The significance of this fiscal weapon should certainly not be underestimated; indeed, the authors of the Minority Report of 1909 described the audit as 'one of the most important features of the new Poor Law'.[20] Much of the correspondence between central and local authorities after 1844 was to be concerned with the reports of auditors; in the late 1850s, for example, the 54 district auditors were estimated to have inspected a total of 100,000 separate books annually.[21] This provided an important control on local expenditure and, indirectly, local relief policy. Its significance cannot be measured simply by the number of disallowances, still less of surcharges actually made,[22] for this would be to ignore its deterrent effects. The audit operated as a constant reminder to Guardians that their decisions were not autonomous. Even in the most minute details of workhouse architecture, for example, the sanction of the central authority was necessary before a proposed expenditure was legal; the threat of surcharge was enough to ensure that Guardians took this power seriously.

Thus far, we have confined our consideration of the administrative impact of 1834 to the level of central authority. However, the changes wrought at local level were just as significant. If the reformed system was to offer any prospect of a revolution in Poor Law government, argued the reformers, then it was necessary to redraw the map of local administration. Here they took their cue from Jeremy Bentham, who had portrayed the confused geography of local government as a fundamental obstacle to rational social reform:

> Looking at the [existing] parochial divisions, at this and that and t'other parish, begotten by chance in the night of darkest antiquity, I see them as an aggregate of heterogeneous fragments, essentially incapable of entering as consistent elements into the composition of any tolerably regular or convenient system.[23]

## Fields of observation: the formation of Poor Law Unions

One of the most visible consequences of the 1834 reform was a transformation in the geography of local government. The parish or township,[24] so central to the workings of the old system, was to be replaced by the Union as the fundamental unit of Poor Law administration. Although the parish was retained as the unit of settlement and chargeability,[25] financial control over the rates passed from local overseers to the Guardians of the Union. During the first four years of the new Poor Law, almost 14,000 local administrative areas were amalgamated into 580 Poor Law Unions, containing 80% of the population of England and Wales (Table 3.1; Figures 3.1–3.3). The first Unions to be formed were concentrated in southern England, the supposed site of the most prevalent 'abuses' under the old

Table 3.1. *The process of Union formation, 1835–1838*

| | Total number of unions in existence | Proportion of all parishes under union (%) | Proportion of all population under union (%) |
|---|---|---|---|
| August 1835 | 114 | 14 | 9 |
| August 1836 | 365 | 57 | 42 |
| July 1837 | 564 | 93 | 78 |
| August 1838 | 580 | 95 | 80 |

*Source:* PLC, *Annual Reports*

*Note:* This table is based on figures abstracted from the Appendices to the Poor Law Commissioners' Annual Reports. The proportions in the second column refer to all places which raised a separate poor rate in 1831 (including townships), and those in the third column are based on population totals in the 1831 Census.

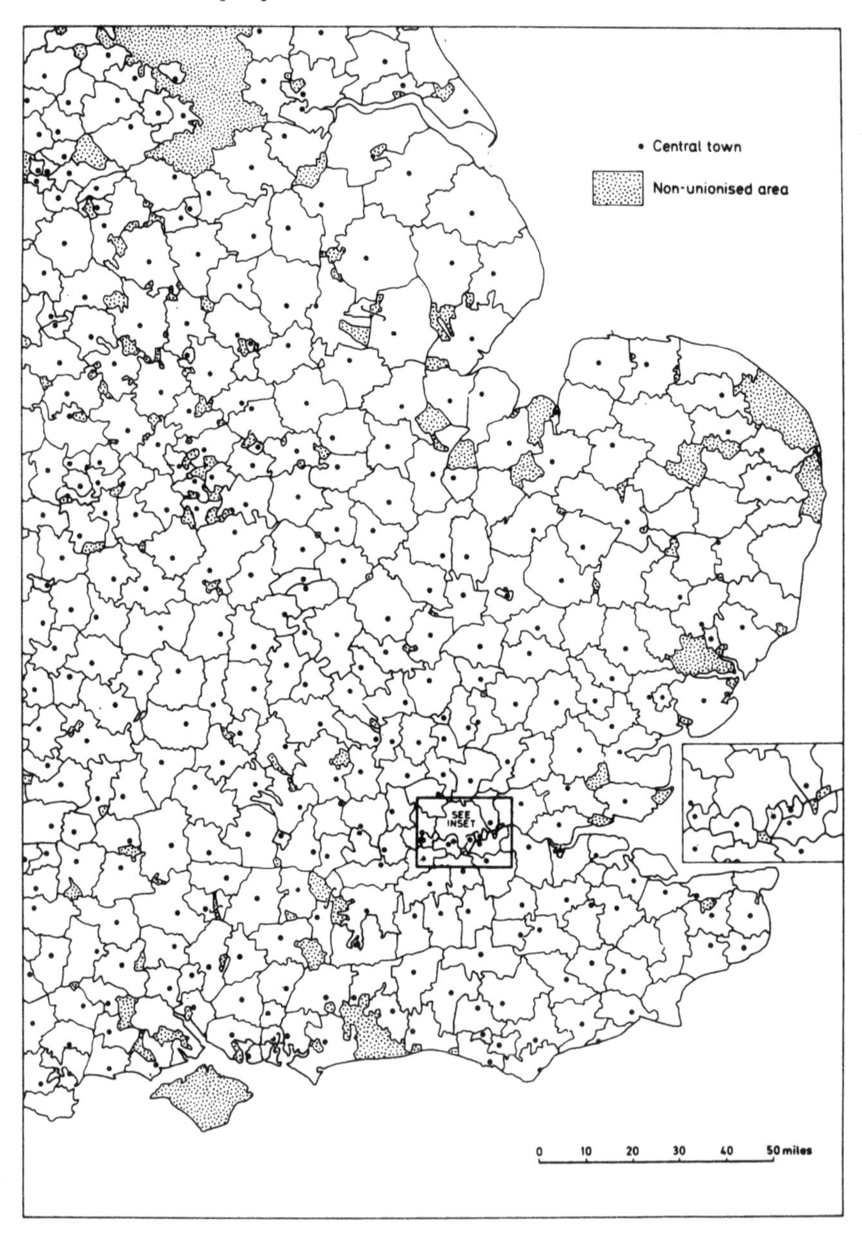


Figure 3.1 Poor Law Unions in the South and East, 1846 (Compiled and redrawn from E. Pierce, 'Town–country relations in England and Wales in the pre-railway age', Unpublished M.A. thesis, University of London, 1957)

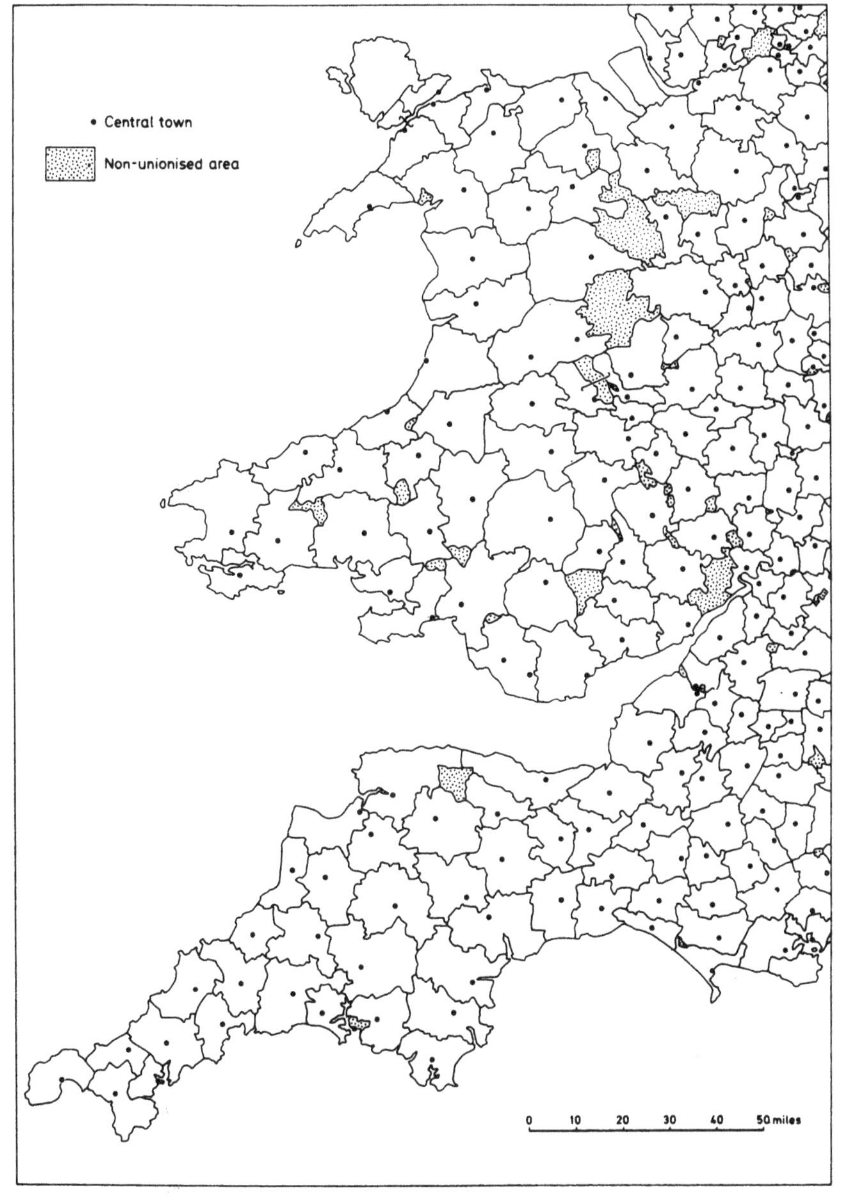


Figure 3.2 Poor Law Unions in the South and West, 1846 (*Source*: see Figure 3.1)

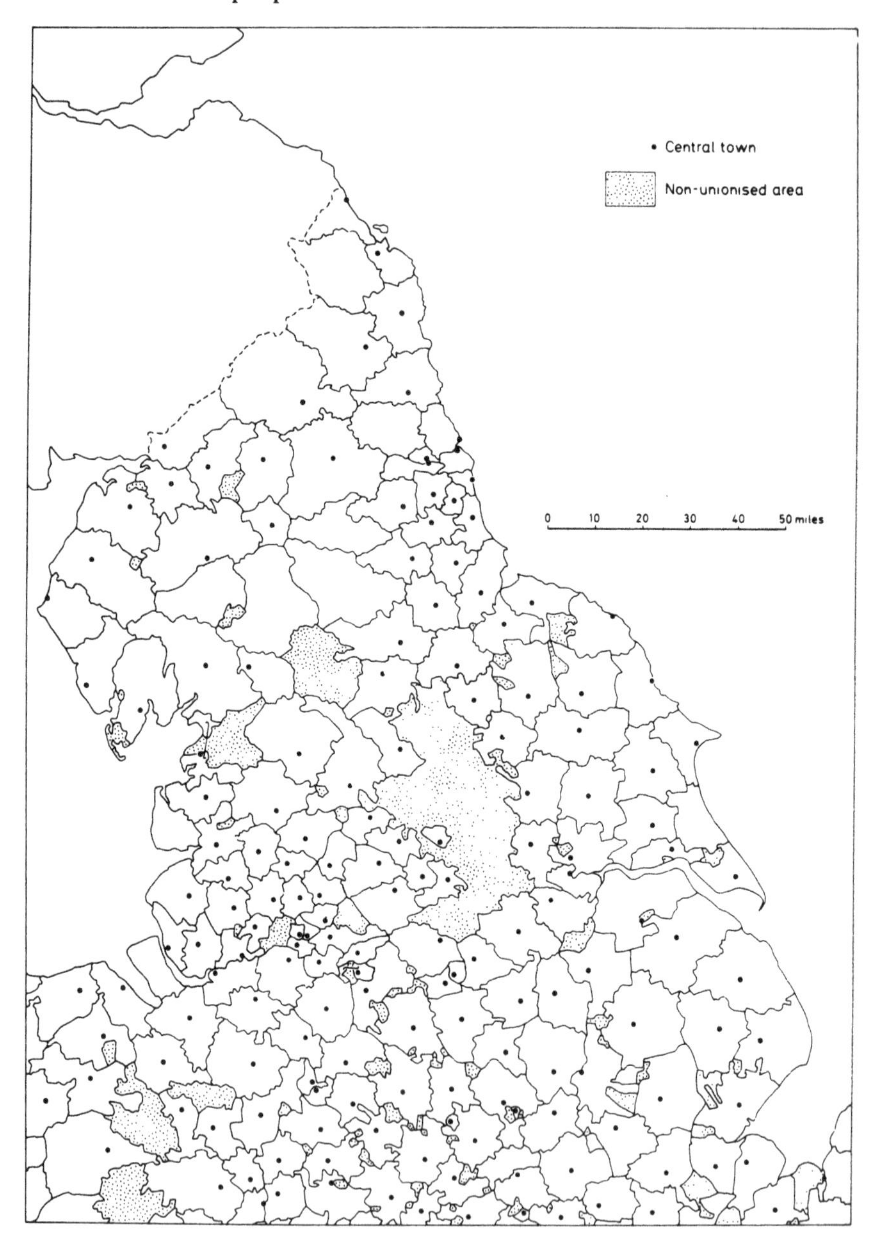


Figure 3.3 Poor Law Unions in the Midlands and North, 1846 (*Source*: see Figure 3.1)

system. Subsequently, the central authority turned its attention to London, the Midlands and Wales (in 1836), and the industrial North (in 1837). While the transformation of these years was to be strategically incomplete, owing to the survival of incorporations formed before 1834 (discussed in the following section), the establishment of a national system of Poor Law Unions clearly inaugurated a new era of local administration.

The formation of Poor Law Unions has frequently been interpreted in terms of economic efficiency, as a step in the 'rationalisation' of the geography of government.[26] The architects of the new system themselves described the process in this way, insisting that Poor Law Unions would provide 'a more efficient machinery for the administration of the law, ... at a less cost and with greater certainty than any other course of proceeding'.[27] In the face of considerable scepticism, Edwin Chadwick insisted that just as the smallest parishes before 1834 had been the most heavily pauperised, so the largest Unions after 1834 would achieve the greatest savings.[28] Amalgamation into Unions was designed to allow local authorities to capitalise on the advantages of scale, not only through a more efficient administration, but by appointing a higher class of local official and by building larger institutions for a greater number of indoor paupers. Yet there was more to it than narrow financial advantage. The new system was intended to reduce the possibility of undue local influence of one kind or another. The authors of the 1834 Report had emphasised the need to remove decisions over relief as far away as possible from the 'door of the pauper'. The move from parish to Union was thus designed to make administration less amenable to undue popular or individual pressure. The parish was dismissed as a 'narrow field of observation', in which judgement was inevitably 'warped by private interests'.[29]

It should by now be clear that the construction of Poor Law Unions was as much a political task as an economic one. If efficiency was the motive, this did not exclude considerations of control and authority. To an extent, this is confirmed by the evidence we have concerning the designation of Union boundaries. In their first Annual Report, the new central Poor Law Commission illustrated the process of Union construction by referring to the drawing of circles around convenient central market towns. The maximum radius of these circles was to be governed by two considerations; one broadly economic (the limits of accessibility), the other broadly political (the need to ensure effective control).[30] This scheme should not be taken too literally; indeed, the new Assistant Poor Law Commissioners were specifically requested to consult local 'interested parties', including large landowners and parochial authorities, before drawing up boundaries. In some areas, like Northamptonshire, the owners and agents of great estates clearly had more than a marginal role in the drawing of new boundaries;[31] in others, their role was less direct. In general terms, the process of Union formation was regionally variable, the outcome depending on such factors as the amount of time available, the influence of local landowners and the pressure of popular agitation.[32]

In the industrial districts of Lancashire, Yorkshire and Cheshire, where the new Poor Law was not introduced until 1837, Unions were formed more hastily than had been the case elsewhere. Alfred Power, the first Assistant Commissioner appointed to these districts (in November 1836) was under considerable pressure to introduce the law as quickly as possible; in any case, he was far from enthusiastic about local consultation. Soon after his assignment to the district, he reported that time was too short to undertake the kind of detailed, local investigations which had been carried out elsewhere prior to the formation of Unions. His plan of Union boundaries seems to have been made without widespread consultation, largely on the basis of administrative convenience.[33] An important consideration in this context was the growing pressure of anti-Poor Law feeling in the North; during his very first local visits, Power found himself mobbed, abused and chased from several towns. The extent of his burden and the speed of the initial implementation in the industrial North was to cause problems in the long run. Charles Mott, a second Assistant Commissioner drafted into the district in August 1838, reported privately to London that local resentment at the rapid process of Union formation in the North had undermined the support of even those sympathetic to the new system.[34]

The historical geography of Union formation thus clearly involved political considerations, although the shape these influences took varied from region to region. The Northern district, for example, was simply too large and too volatile in 1837 for the kind of controlled implementation which had been achieved in the agricultural South. In the long run, however, it was not the anti-Poor Law movement which was to cause the greatest obstacle to the construction of a national system of Unions; a more lasting problem was the survival of the old Poor Law incorporations.

## Lagoons of irrationality: the old Poor Law incorporations

The 1834 Act did not give the Commissioners the power to dissolve existing incorporations formed by Act of Parliament before 1834. As Figure 3.4 indicates, these incorporations covered a large area of land, containing (in 1844) about 10% of the population of England and Wales.[35] The Webbs described them as 'unreformed lagoons of independent administration', preventing the completion of the Union system and interfering with the implementation of the new law in neighbouring parishes.[36] This was certainly a view shared by the central authorities after 1834; and yet only in 1869 did Parliament grant them the unqualified power to dissolve these incorporations without local consent.[37]

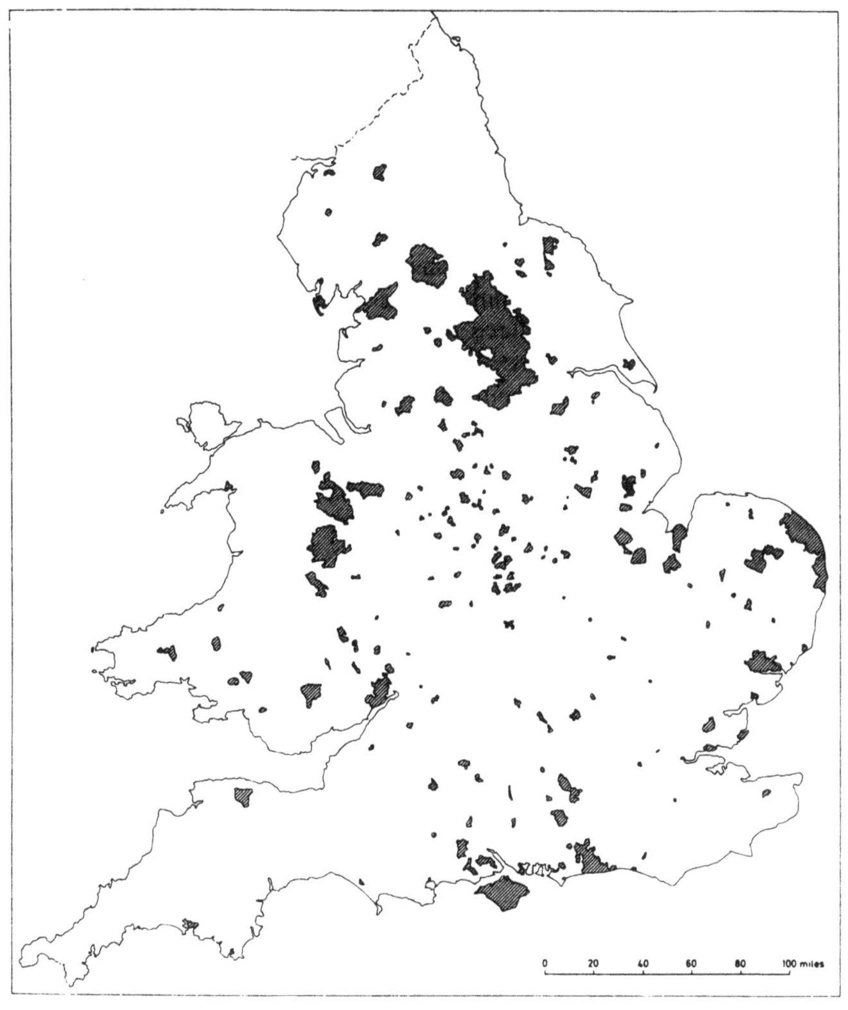


Figure 3.4 Areas not incorporated within a Poor Law Union, 1846 (*Source*: see Figure 3.1)

Although the parish constituted the prime administrative unit under the old Poor Law, the amalgamation of parishes was possible, either under special Local Acts or under Thomas Gilbert's Act of 1782. These two kinds of Union were treated somewhat differently by the central authority after 1834. Approximately 125 incorporations were created under Local Acts between 1647 and 1833.[38] Most, but not all, consisted of several parishes in combination with each other. Many, like the celebrated incorporation at Bristol, ran workhouses of their own; some also took responsibility for out-relief, although in rural areas this seems to have been left largely to the parish authorities. While the Webbs described their performance as almost uniformly 'dismal',[39] it is likely that there was considerable diversity within this large group of incorporations. The Poor Law Commissioners found it difficult to prove widespread mismanagement after 1834, though not for want of trying. During the late 1830s, the Local Act incorporations were repeatedly condemned, in Benthamite terms, as founded on a '*lex loci*, differing from the general system, incapable of assimilation'.[40] But such attacks met with considerable resistance. Although the Commissioners succeeded in obtaining a legal ruling on their right to regulate relief policy in such incorporations, they suffered an embarrassing defeat in their attempt to force through elections under the 1834 Act.[41] During the 1840s, official statements on the matter became more circumspect, with more attention paid to the similarities rather than the differences in relief policy between Poor Law Unions and Local Act Unions.[42] This strategy reflects the more general drift to a less aggressive official strategy during the 1840s; but it also clearly reflected the strategic location and relative power of the Local Act Unions. They were characteristically urban rather than rural, including for example such places as Birmingham, Liverpool, Norwich and many of the London parishes (Figure 3.5). After the mid 1840s, the central authority seems to have resigned itself to the continued existence of the Local Act Unions at least until Parliament granted it powers of dissolution.

An alternative form of incorporation under the old Poor Law was provided by Gilbert's Act of 1782. This enabled parishes to unite for the maintenance of a central poorhouse, with the consent of two-thirds of the ratepayers. The Act was described by its sponsor, Thomas Gilbert, as a small step towards 'the grand principle of *Union, Superintendence, Controul and Permanency of Office*'.[43] A Union of parishes, Gilbert argued, would facilitate the construction of efficient institutions and the appointment of a higher class of paid and permanent staff. Sympathetic to these arguments, the Webbs described the 1782 Act which bears his name as 'the most carefully devised, the most elaborate and perhaps the most influential, for both good and evil, of all the scores of Poor Law statutes between 1601 and 1834'.[44] In view of this recommendation, it is perhaps surprising that no study of the implementation of Gilbert's Act yet exists. It seems that the Act was not immediately popular, although there was a slow and steady increase in the number of Gilbert's Unions formed during the early nineteenth century. By 1834, at least sixty-eight Unions and three single parishes had adopted the provisions of the Act, incorporating in total one thousand parishes containing half a million people.[45] In 1850, the number of Unions and single parishes adopting the Act had declined to twelve and two respectively, the population within them (less than two hundred thousand) representing a small fraction of those still incorporated within Local Act Unions.[46]

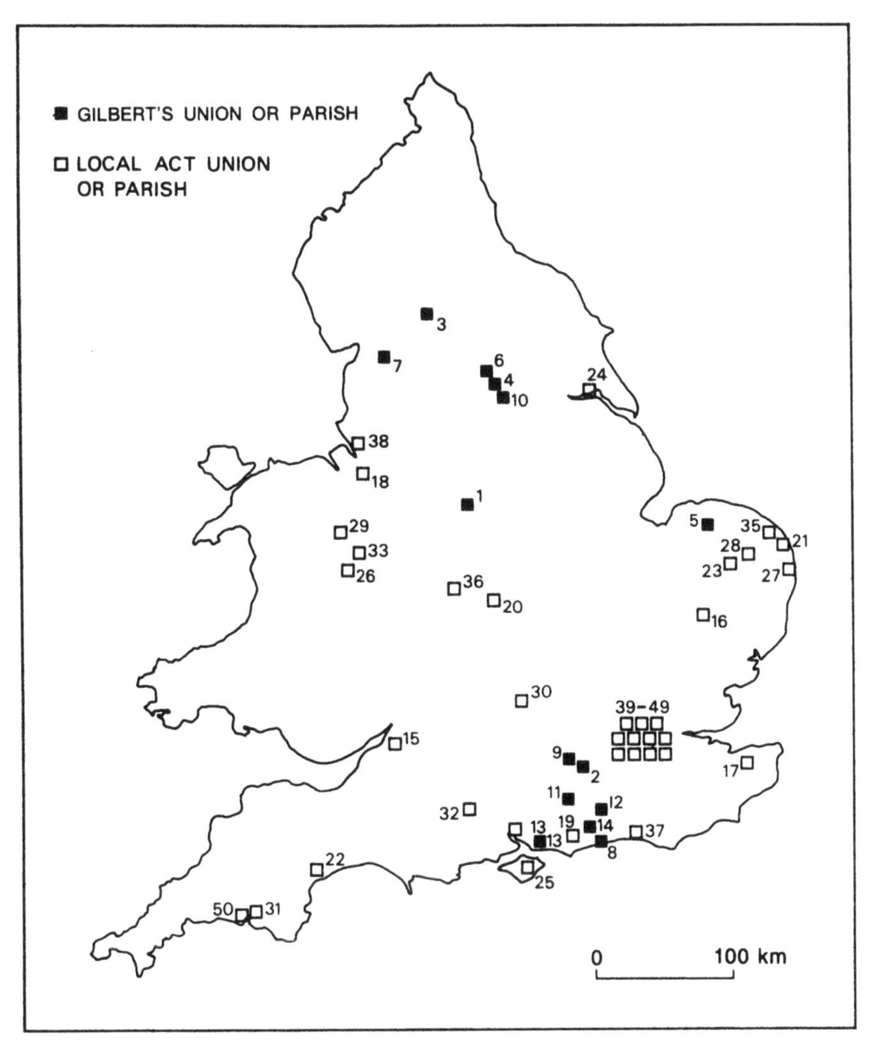


Figure 3.5 Gilbert and Local Act Unions in England and Wales, 1856 (*Source*: Parliamentary Return, PP 1856 XLIX)

*Gilbert's Unions*
1. Alstonefield
2. Ash
3. Bainbridge
4. Barwick
5. Brinton
6. Carlton
7. Caton
8. East Preston
9. Farnborough
10. Great Preston
11. Headley
12. Sutton

*Gilbert's parishes*
13. Alverstoke
14. Arundel

*Local Act Unions*
15. Bristol
16. Bury St Edmunds
17. Canterbury
18. Chester
19. Chichester
20. Coventry
21. East & West Flegg
22. Exeter
23. Forehoe
24. Hull
25. Isle of Wight
26. Montgomery & Pool
27. Mutford & Lothingland
28. Norwich
29. Oswestry
30. Oxford
31. Plymouth
32. Salisbury
33. Shrewsbury
34. Southampton
35. Tunstead & Happing

*Local Act parishes*
36. Birmingham
37. Brighton
38. Liverpool
39. St George Hanover Sq
40. St Giles & St George
41. St James Clerkenwell
42. St James Westminster
43. St Leonard Shoreditch
44. St Luke Middlesex
45. St Margaret & St John
46. St Mary Islington
47. St Marylebone
48. St Mary Newington
49. St Pancras
50. Stoke Damerell

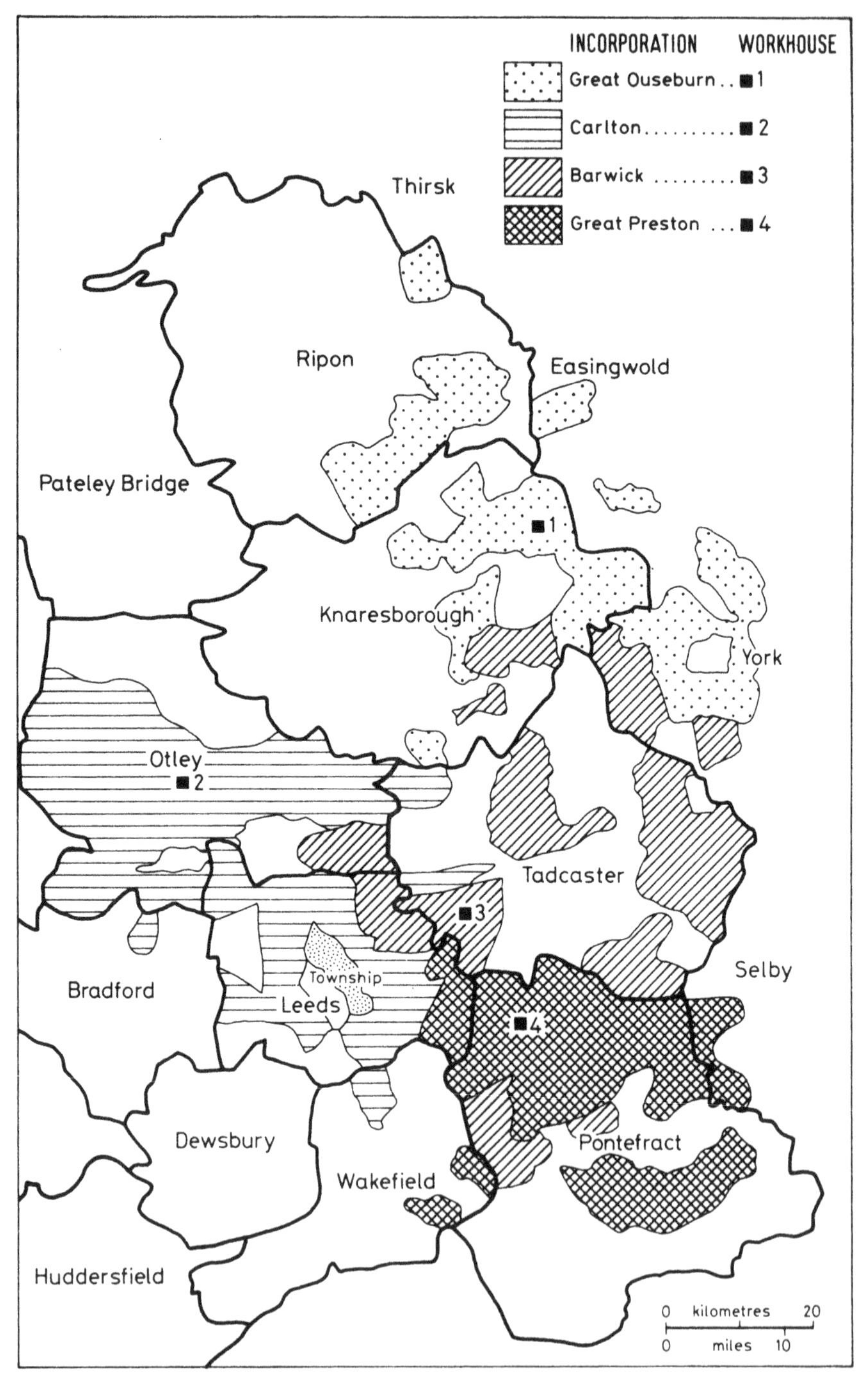


Figure 3.6 Gilbert Unions in the West Riding of Yorkshire, 1837 (*Source*: S. C. Poor Law, P.P. 1837–8 XVIII). The proposed new Poor Law Unions are also shown.

There is a remarkable similarity between Thomas Gilbert's arguments for the incorporation of parishes and Edwin Chadwick's case for the formation of Unions in 1834. (In the light of Gilbert's more positive reputation amongst Poor Law historians, it is interesting to note that the administrative aspect of his reforms attracted similar criticisms to those levelled against the 1834 reform.)[47] Nevertheless, the architects of the new Poor Law themselves saw two fatal flaws in Gilbert's Act. Firstly, they argued, it failed to deter the able-bodied from applying for out-door relief, since its workhouses were reserved for the aged and infirm, and there was little apparent restriction on the granting of relief by individual Guardians.[48] Secondly, it created Unions on an entirely *ad hoc* basis, without reference to contiguity or other geographical considerations, beyond the requirement that the common workhouse had by law to be within ten miles of each individual parish. This lack of contiguity caused the Commissioners considerable difficulty in drawing up boundaries for Poor Law Unions in areas adjacent to Gilbert's Union parishes. Indeed, even the most ardent supporter of Gilbert's Act conceded (in 1841) that 'the map is conclusively against us'.[49] Figure 3.6, for example, shows the location of 162 individual townships (containing 100,000 people) incorporated within four Gilbert's Unions in the West Riding of Yorkshire. Alfred Power, the Assistant Commissioner responsible for the district, proposed (unsuccessfully) to replace the existing Gilbert's Unions with six 'regularly formed' Unions, centred on the most important market towns.[50]

After 1834, the right of the central authority to intervene in the running of the Gilbert's Unions was the subject of a series of running battles. The Commissioners insisted that many of their powers were enforceable in Gilbert's Unions, as well as those formed under the new Poor Law. Yet, in 1840, even their right to request a statistical return was successfully challenged at the court of Queen's Bench.[51] This represented an important victory for the champions of local autonomy. Despite further attempts during the 1840s, the Commissioners failed to secure Parliamentary approval for the dissolution of all but the smallest of Gilbert's Unions. Once the issue was described in the language of local rights, the central authority frequently found its plans frustrated by powerful opposition. As one M.P. put it in 1844,

> The Commissioners represented the Gilbert Unions as nuisances – not so much to the public, as to Somerset House, being geographically awkward for the purposes of Poor Law divisions: but it was too bad that for this reason ... they were to be set aside.[52]

### The quest for uniformity: relief policy on the map

The architects of the new Poor Law argued for greater uniformity in the administration of relief. It is important to recognise, however, that this quest for 'uniformity' had several dimensions. One, as we have seen, was administrative; it required the standardisation of local procedures. Another was concerned with policy; it required local relief practice to be founded on principles consistent with those of the central authority. Yet another kind of uniformity – the levelling out of variations in pauperism – was often implied but not explicitly advocated in 1834. In general, then, the reform was designed not to eradicate local differences, but to make them more *manageable*.

Although the central authority had considerable influence over the administration of the Poor Law, its powers in the sphere of out-door relief policy were in fact quite limited. After 1834, the Commissioners developed three formal mechanisms for the regulation of local relief practice, all largely confined to the able-bodied. The *Prohibitory Order* stipulated that all able-bodied paupers should be relieved only inside workhouses, unless they could be considered exceptional cases; the latter included 'cases of sudden or urgent necessity', cases of 'sickness, accident or bodily or mental infirmity', widows with children and the bereaved. This Order, as we shall see, was far from universally applied; in some places, it was relaxed by a Supplementary *Labour Test Order*, which allowed out-door relief to able-bodied paupers, in return for a task of labour. A version of this latter Order was issued separately to a large number of Unions in 1852 as the *Relief Regulation Order*. This required that, with certain exceptions, able-bodied male paupers be set to work in return for out-door relief. It also stipulated that at least half of any relief should be given in kind rather than money, at no more than weekly intervals. Relief to able-bodied males already in employment was forbidden.

Historians are divided as to the effect of these Orders. Most have assumed that a combination of local recalcitrance and economic pressures conspired to defeat the relief policies of the new authority. Karel Williams, in contrast, has argued that the reduction in male able-bodied pauperism after 1834 constituted 'a brilliant triumph in official strategy'.[53] While Williams' criticisms of the received wisdom are generally convincing, his own case rests on a relatively narrow definition of central policy, confining it to the abolition of any practice involving the granting of out-door relief on a regular basis to able-bodied men in paid employment. More generally, Williams portrays 'official strategy' as if it were a monolithic programme, the direct product of the 1834 Report, undifferentiated at the centre and virtually unopposed in the localities. The evidence suggests, however, that the 1834 Report did not pre-determine the form or the outcome of subsequent official policy. Indeed, the historical geography of subsequent relief policy reflects the impact of struggles which were endemic within the new administrative system.

Central strategy on relief policy appeared straightforward enough in 1835; in their first Annual Report, the Commissioners happily contemplated 'the complete adoption of the workhouse system and the extinction of all out-door relief to the able-bodied'.[54] But in their very next Report, it was announced that the application of this policy to women had been 'deferred'.[55] By 1837, the Commissioners' relief policies were coming under increasing attack. The mass anti-Poor Law campaign was gathering pace, with violent crowds greeting Assistant Commissioner Power as he attempted to implement the law in towns such as Huddersfield and Bradford. The Commissioners were undoubtedly under great pressure during this period; Frankland Lewis complained to John Russell, the Home Secretary, that 'the peculiarly anxious and responsible character of the business of the office – the constant obloquy which is so perseveringly poured out on all those who administer it ... has already produced [a change] in my whole condition of existence'.[56] These were hardly ideal conditions for consistent and far-sighted policy-making. Russell himself, under pressure from Whigs in the industrial North, advised delay and caution in the face of popular opposition and a deepening economic crisis[57] (see pp. 121–2). The result can only be seen as a tactical retreat.

In their fourth Annual Report, published in August 1838, the Commissioners announced a major concession to the 'peculiar circumstances' of Lancashire and the West Riding of Yorkshire. Owing to the unusually high level of casual relief and settlement business in these areas (and, we might add, the political impact of anti-Poor Law protest), a greater degree of local autonomy in the appointment of officials was to be permitted in the industrial North. Moreover, there were to be no Prohibitory Orders regulating out-door relief; it would instead be left to the discretion of the Guardians 'to realise the objects of those rules in such a manner and to such an extent as they might find compatible with the circumstances of their respective Unions'.[58] Such a retreat was unprecedented. It appeared to hold out the prospect of a permanent concession to the 'peculiarities' of the industrial North. Elsewhere, there were further changes. In other urban Unions such as Nottingham, Prohibitory Orders were relaxed through the addition of Labour Test Orders. In an attempt to justify this retreat from the policy applied elsewhere, the Commissioners suggested that 'uniformity of principle is incompatible with uniformity in the administrative details, unless (which is impossible) the local circumstances be everywhere identical'.[59] However, by their own admission, the labour test was a poor substitute for the workhouse test. In 1835, the Commissioners had themselves gone to some lengths to discredit the idea of a 'labour test', arguing that it failed to draw an adequate line between paupers and independent labourers.[60] The formal codification of the labour test undermined the credibility of the Commissioners' workhouse-based strategy. According to many of their own officials, the policies of the Commissioners after 1838 represented a betrayal of the 'principles' of 1834. As Edward Tufnell, a senior workhouse schools inspector, complained in 1847, 'their sole object for the past few years seems to have been to endeavour to cull popularity by acting the "poor man's friend"' [61]

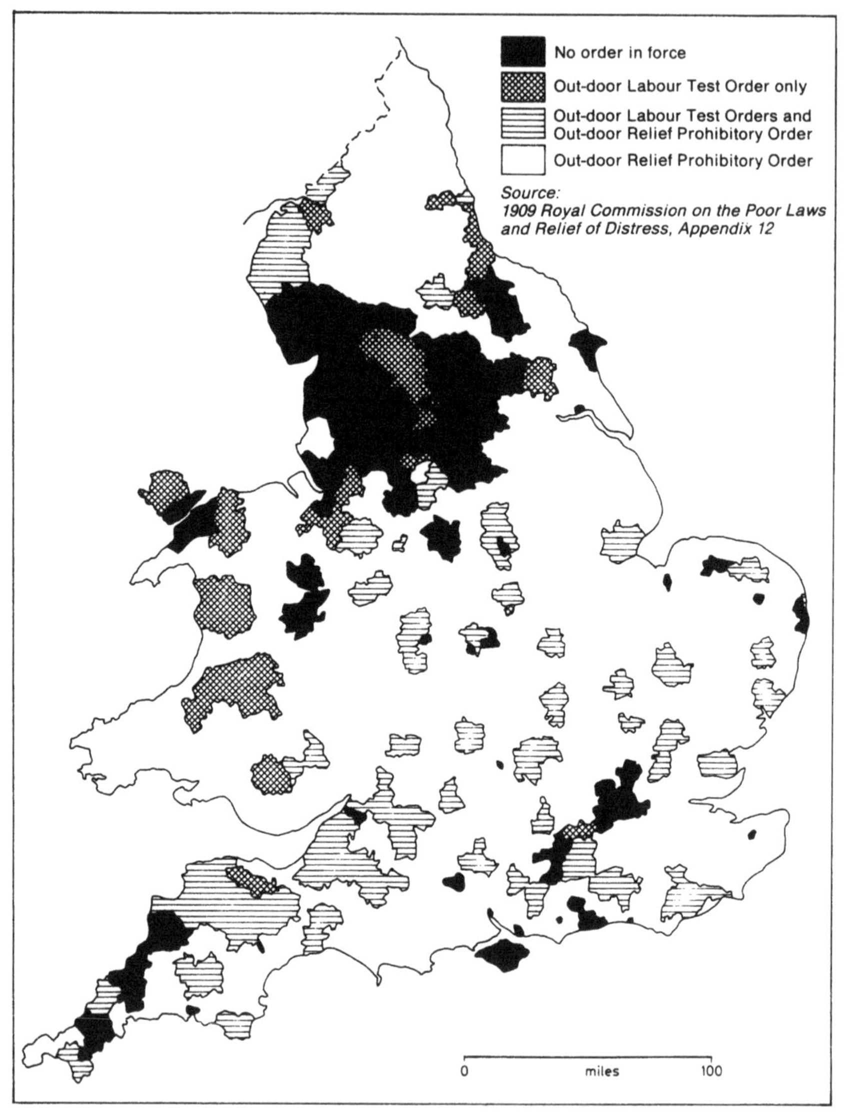


Figure 3.7 Out-door relief to the able-bodied: the geography of regulation, 1847 (*Source*: R. C. Poor Laws, P.P. 1910 LI)

As Figure 3.7 shows, the impact of central relief regulation during the 1840s was regionally variable. In many parts of England and Wales, a Labour Test Order was in operation in 1847; and in large parts of the industrial North and London, as well as in the surviving old Poor Law incorporations, there was no Order in force at all. This situation was altered

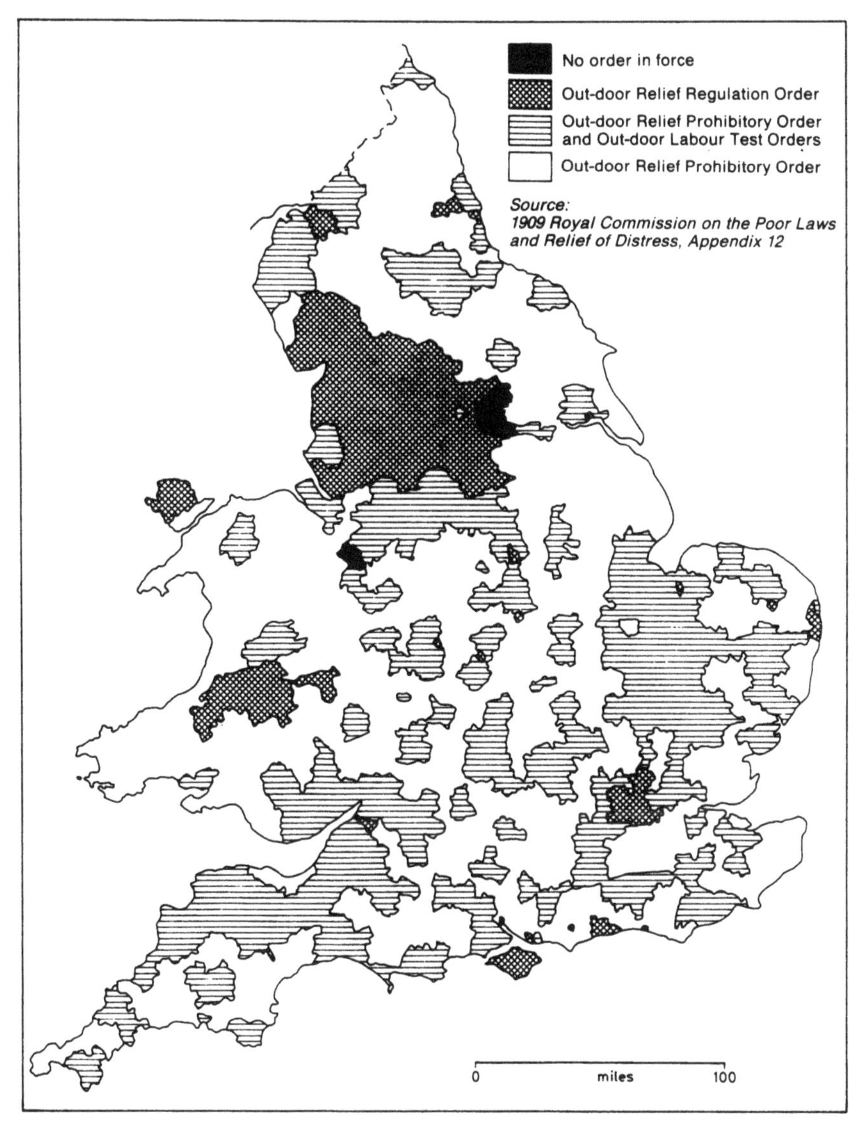


Figure 3.8 Out-door relief to the able-bodied: the geography of regulation, 1871 (*Source*: see Figure 3.7)

in 1852, with the issue of an Out-door Relief Regulation Order to many of the industrial and urban Unions. The original version of this Order provoked fierce opposition from Boards of Guardians in the industrial North and in London. At stake were competing interpretations of local autonomy as well as questions of relief policy. After a surprisingly short and well-

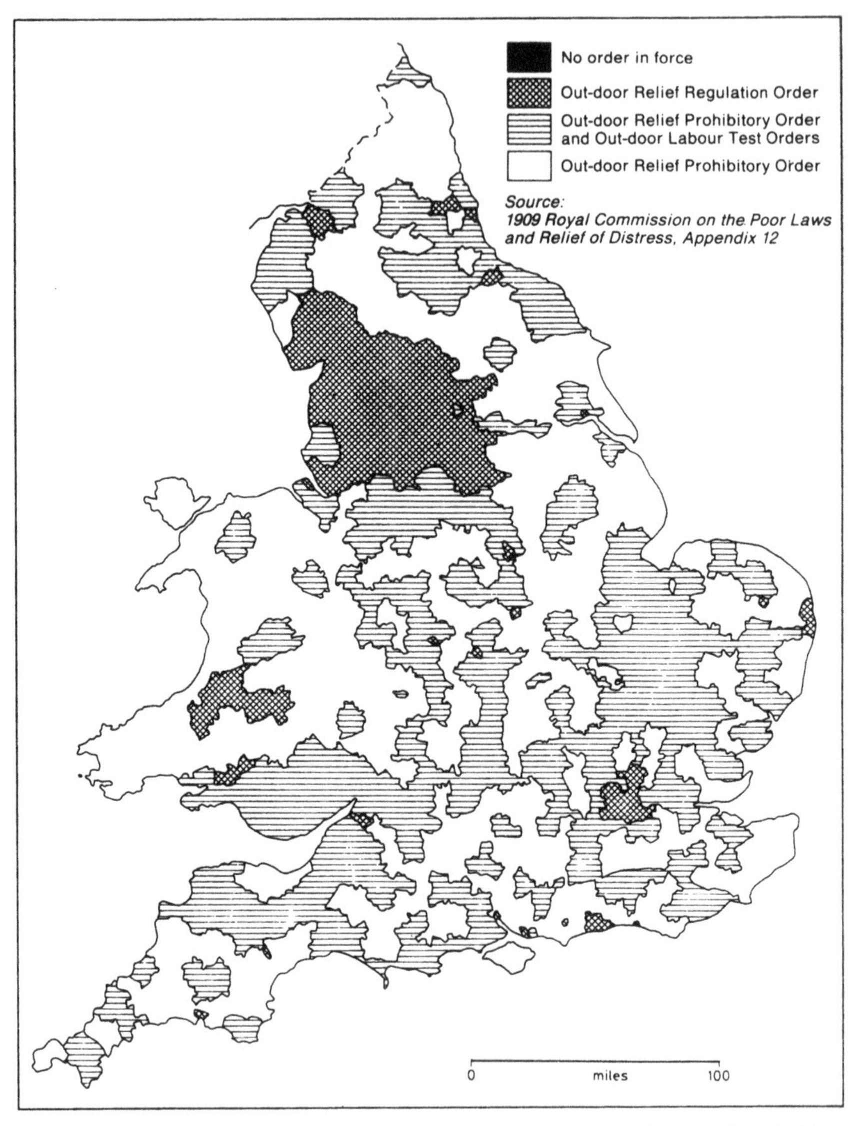


Figure 3.9 Out-door relief to the able-bodied: the geography of regulation, 1906 (*Source*: see Figure 3.7)

orchestrated campaign, the central authority agreed to revise its Order. The new version (unlike its predecessor) gave the Guardians full discretion in their policy towards the non-able-bodied, and allowed them more autonomy in the granting of short-term relief to the able-bodied. More significantly, the Poor Law Board advised that the exclusion of relief to men in employment applied only when relief was given 'at the same identical time as that at which the person receiving it is in actual employment and in receipt of wages'. The Board also emphasised its willingness to suspend the Order in times of distress.[62]

The struggle over the 1852 Order provides another instance of the revision of central policy in the face of local opposition. Rather than simply finishing off the job of regulation, as Williams suggests,[63] the 1852 Order institutionalised a compromise. As Figures 3.8 and 3.9 indicate, the Order remained in force in the industrial North, in London and in central Wales throughout the nineteenth century. Elsewhere, the Prohibitory Order appears to have been relaxed still further, with Labour Test Orders in operation in many parts of the country by the turn of the century. The overall impression given by these Figures is one of organised diversity. The extension of central relief regulation was achieved at the price of important concessions to local Guardians, particularly in urban and industrial Unions. Many of the supporters of the new Poor Law (including Edwin Chadwick) regarded such concessions as fatal to the success of the 'principles' of 1834. Even the diplomatic Charles Mott was critical of the retreats of the late 1830s; and several central Poor Law inspectors regarded the alterations to the 1852 Order as instrumental in the continuation of practices (including relief-in-aid-of-wages) contrary to the 'first principles' of the 1834 law.[64]

After the controversies of 1852, the Poor Law Board abstained from further formal revisions of relief policy. To some extent, the Board simply lost the initiative during the 1850s and 1860s; indeed, it repeatedly found itself the object of criticism from professional and specialist agencies advocating workhouse reform. The late 1860s witnessed a revival of concern over out-door relief, promoted by groups such as the Charity Organisation Society. What followed was an intensification of the administrative war against out-door relief in general; those Unions with a low proportion of their paupers on out-door relief were simply considered to be efficient, those with higher proportions were considered lax. In 1872, the Local Government Board issued tables showing the ratio of out-door to indoor pauperism in each Union, portraying them as a straightforward guide to the geography of administrative efficiency. Figure 3.10 shows those Unions where the ratio was highest and those where it was lowest, respectively. It highlights some remarkable contrasts in relief practice between Unions experiencing similar levels of overall pauperism (cf. Figure 3.11). For example, a block of Unions in East Lancashire and Cheshire (including Ashton, Manchester and Salford) recorded high proportions of indoor pauperism, while neighbouring Unions in the West Riding (including Huddersfield, Dewsbury and Wakefield) maintained far fewer of their paupers in workhouses, without suffering significantly higher levels of overall pauperism. Nevertheless, the central authority continued to insist on their correlation between high rates of indoor

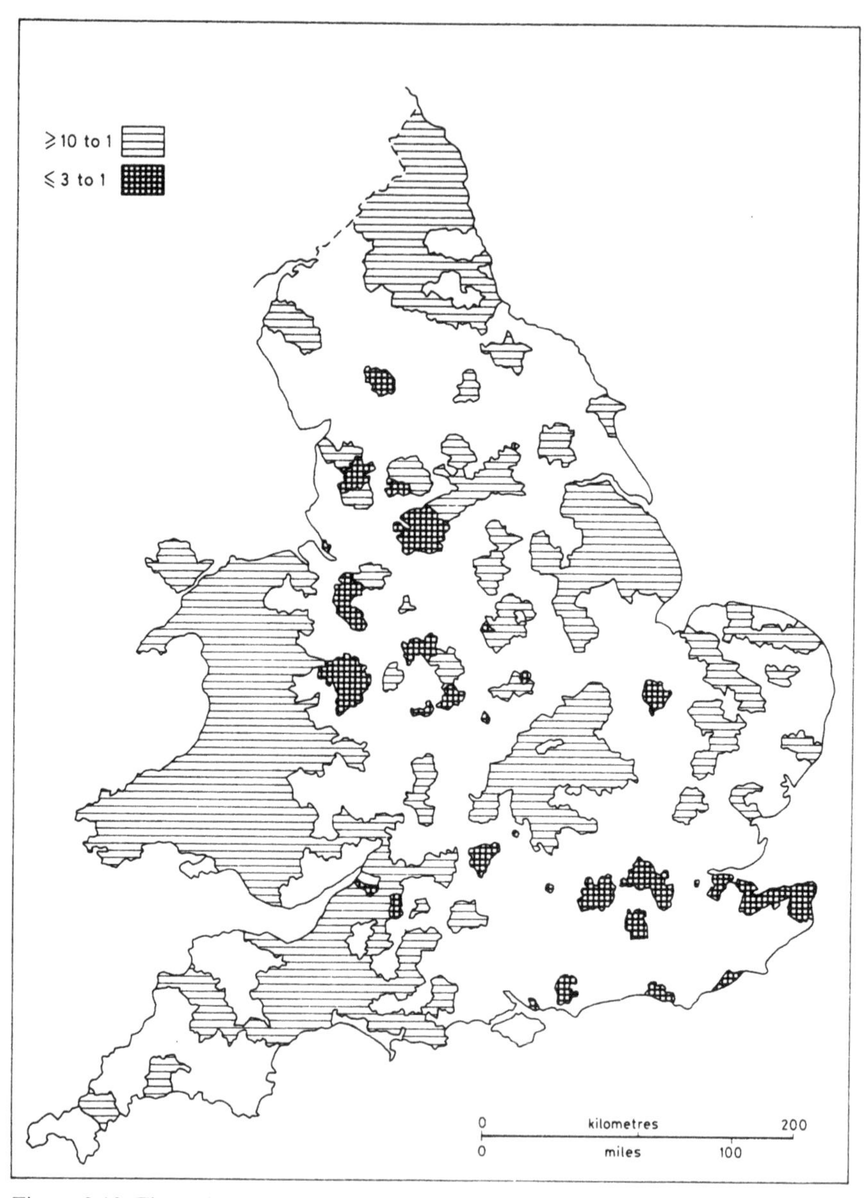


Figure 3.10 The ratio of out-door pauperism to indoor pauperism, by Union, 1872 (*Source*: LGB, *Second Annual Report*, 1873, Appendix D65)

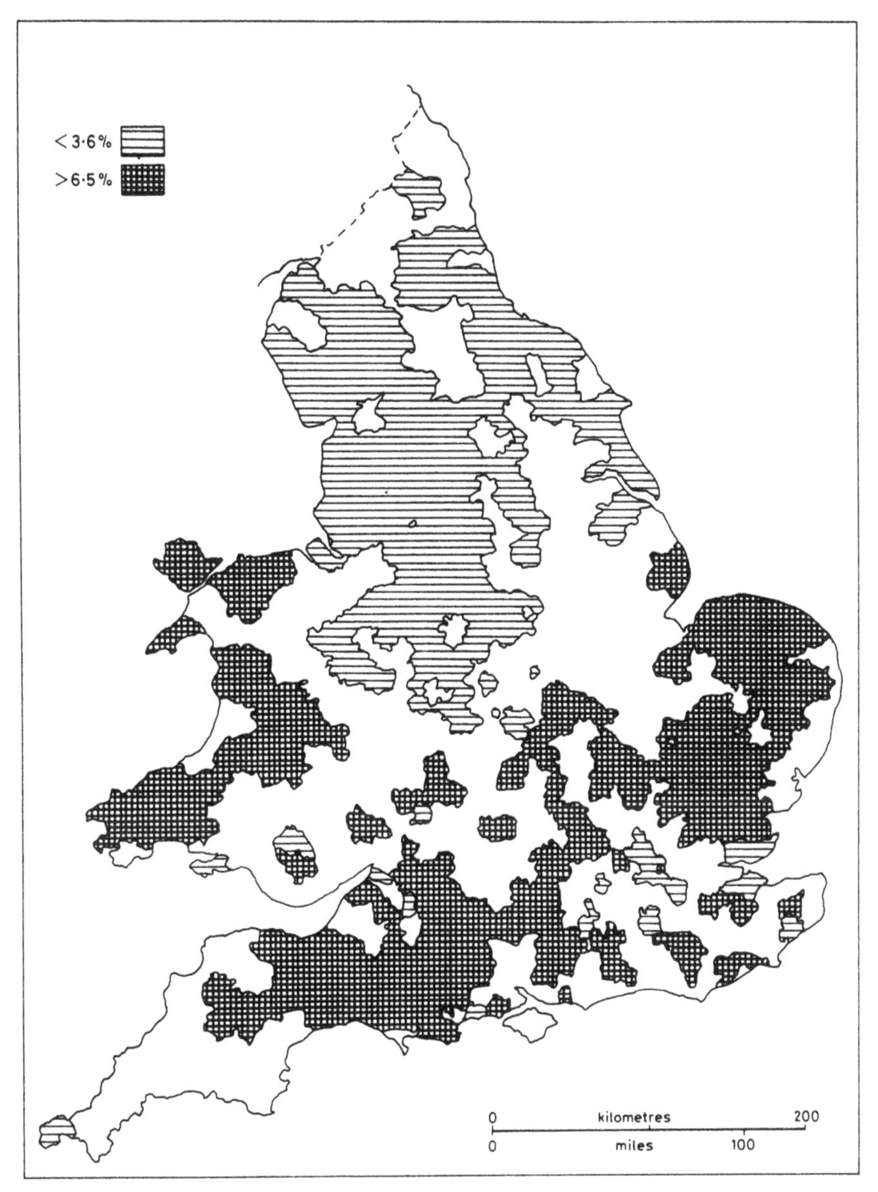


Figure 3.11 The ratio of pauperism to population, by Union, 1872 (*Source*: see Figure 3.10)

relief and low rates of overall pauperism. During the 1870s, many Boards of Guardians introduced stricter rules on the eligibility of paupers for out-door relief. In London, where administrative reorganisation permitted a vast expansion of institutional provision, particularly dramatic reductions in out-door relief were achieved; the proportion of paupers receiving out-door relief in the Metropolis fell from over 70% in January 1872 to 45% in 1885, and was less than 30% in 1895.[65]

Throughout the 'crusade' against out-door relief, the Local Government Board eschewed formal measures of relief regulation in the shape of new Orders, relying instead on a combination of official persuasion, local initiatives and pressure group activism.[66] This strategy provides one indication of a more general shift in the *modus operandi* of the central authority since the 1830s; indeed, one historian describes the Local Government Board 'not as a leading edge of the instrumental state, but a byword for compromise, pragmatism and conservatism'.[67] Although such a description seems even more appropriate in the case of its predecessor, at least as far as relief policy is concerned, the Local Government Board remained cautious in its approach to local Boards of Guardians. In preferring the flexibility of the advisory circular to the bluntness of the mandatory Order, the Board distanced itself from the more abrasive approach adopted by the Poor Law Commissioners during the 1830s. However, the effect was a further weakening of its attempts to promote uniformity in relief practice. By the end of the century, a plethora of rules and regulations governed the administration of out-door relief in the Unions of England and Wales. This diversity was to prove an easy target for critics of the central authority; the Minority Report of 1909, for example, interpreted the map of relief regulation as evidence of a 'flagrant disregard' for uniformity, one of the cardinal principles of 1834.[68]

## Conclusion

The 1834 Act created a new landscape of Poor Law government, replacing a disjointed network of dispersed and autonomous localities with a genuine administrative system. The new Unions created in 1834 were subsequently to be described as 'administrative communities of the highest importance',[69] in recognition of the functions they fulfilled in this and other areas of government. The construction of a new administrative geography was integral to the reformers' case in 1834, not only because it clearly inscribed the principle of uniformity on the very map of Poor Law government, but also because it enabled the more efficient transmission of standard procedures and practices. I say enabled, but not dictated; for the reformers were clearly unable to determine, in advance, subsequent patterns of relief regulation. The task of actually managing the new system was to be fraught with conflict; as we have seen, compromise was etched onto the map of relief policy. In the following chapters, we turn to the workhouse system itself. Here, as in other spheres of Poor Law administration, the 1834 reform set the stage for a dramatic shift in policy; but it could not dictate the script of subsequent practice.

# 4

# Designing the workhouse system, 1834–1884

In November 1834, the Poor Law Commissioners described the workhouse system as 'the only remedy which can be entirely depended upon for the mitigation and ultimate extinction of the various evils which have been generated through the faulty administration of the Poor Laws'.[1] This claim, so typical of the single-minded spirit of the reformers, provided the basis for an entirely new system of institutional provision. The Commissioners' strategy would require a building programme of unprecedented scope, creating in the course of a few years a dense network of Union workhouses throughout England and Wales. Yet the new system did not turn out to be quite the panacea the reformers had hoped for. Controversies over workhouse design and policy were to command the attention of successive generations of social reformers and policy-makers after 1834. Meanwhile, the spectre of 'the Union' would come to haunt the popular imagination in a way unmatched by any other institution. For these reasons alone, the history of workhouse policy illuminates far more than Poor Law historians have sometimes allowed. The 'bastile' provides a vantage point from which to survey a far broader social landscape.

This chapter is concerned with the evolution of official workhouse policy after 1834. It pays particular attention to the reformers' emphasis on strategies of design and the ambiguities surrounding the idea of classification, a concept which informs the writings of contemporary social reformers, from John Howard in the eighteenth century to Florence Nightingale in the nineteenth. (The importance of this concept within debates over workhouse children and the pauper insane is amply demonstrated in chapter 6.) The workhouse strategy of 1834 combined the semiotics of deterrence and the disciplines of institutional regimes – themes introduced in chapter 1 – although official emphasis was initially on the former rather than the latter. The 1860s marked an important moment in the history of official policy, as several historians have pointed out, and a focus on contemporary debate at this time helps to illuminate the general course of workhouse policy during the period as a whole. The changing pattern of actual workhouse provision is traced in chapter 5.

## Strategies of design

Although the Poor Law Commissioners initially seem to have favoured the continued use of a number of existing parish workhouses in each Union, they were soon converted to the view that in most cases the construction of a large new workhouse was essential to the success of their reform. Guardians were therefore encouraged to replace existing institutions with a single Union workhouse as soon as possible after the introduction of the new law. Not only would this be more efficient, argued the Commissioners; it would also be a powerful symbol of an entirely new approach to relief provision. The buildings themselves were designed to make an impression on the poor, as Assistant Commissioner Sir Francis Head explained:

> The very sight of a well-built efficient establishment would give confidence to the Poor Law Guardians; the sight and weekly assemblage of all the servants of their Union would make them feel proud of their office; the appointment of a chaplain would give dignity to the whole arrangement, while the pauper would feel it was utterly impossible to contend against it.[2]

According to another plain-speaking Assistant Commissioner, the forbidding look of the new workhouses was intended as a 'terror to the able-bodied population';[3] yet another remarked in 1836 that 'their prison-like appearance ... inspires a salutary dread of them'.[4] The officially-recommended workhouse plans, drawn up by the Commissioners' young architect Sampson Kempthorne, were later described by George Gilbert Scott as 'a set of ready-made designs of the meanest possible character'.[5] It was widely claimed that Kempthorne had copied his plans directly from designs for American prisons. Even sympathetic commentators readily acknowledged that Kempthorne's 'unhappy designs ... suggested the idea of Bastiles'.[6] It is difficult to imagine a better illustration of the semiotics of deterrence at work.

In their *First Annual Report*, the Commissioners published plans for four model workhouses, one designed by Francis Head, the other three by Sampson Kempthorne. Head's plan for a rural workhouse was intended to house five hundred paupers in a large rectangular block surrounding an open yard, subdivided to separate male and female paupers. Critics dismissed the proposed building as a mere warehouse for the poor, complaining that the special needs of the sick and the young had effectively been sacrificed on the altars of discipline and economy.[7] The few workhouses that were built to this design were certainly large, by English standards; but their size was as nothing compared to those subsequently erected in Ireland,

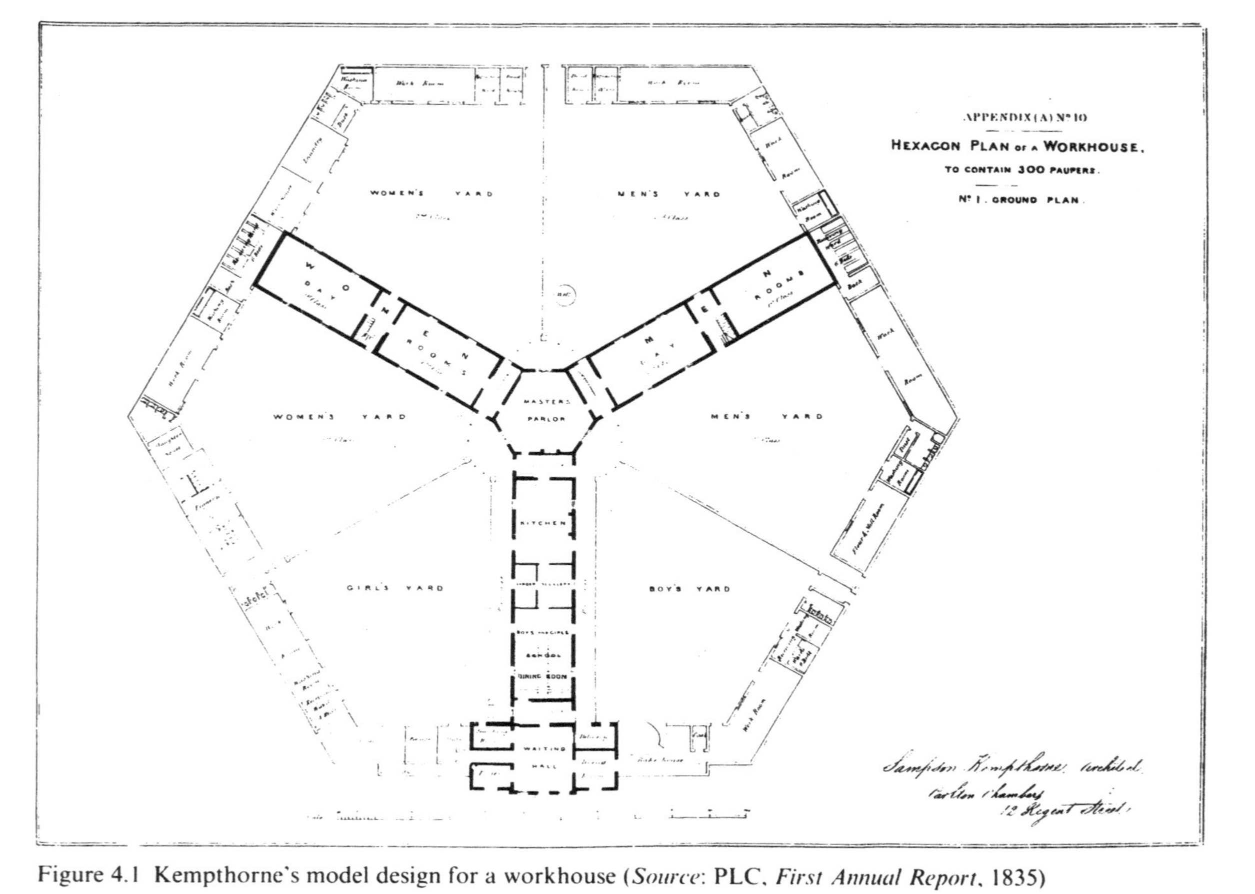


Figure 4.1 Kempthorne's model design for a workhouse (*Source*: PLC, *First Annual Report*, 1835)

where the Commissioners even contemplated converting barracks into workhouses.[8] The Irish workhouses, executed in a style described by one observer as 'a sort of Bastard or Vauxhall Gothic', were in their turn dismissed as 'a series of constantly recurring rural eye-sores, ... their whole aspect affectedly gloomy, narrow and repulsive'.[9] Most of their English counterparts were somewhat smaller, though (to their intended clientele) no less repulsive. Kempthorne's influential designs, for example, catered for between 200 and 500 paupers. They were based on two designs – one cruciform, and the other hexagonal (Figure 4.1) – in which the master's quarters were located at the point of convergence of the accommodation blocks, symbolically reinforcing the centrality of disciplinary order within the new workhouses. Particular attention was also given to the separation of pauper classes, with dayrooms for each sex, separate wards for children and special provision for the sick. Kempthorne's cruciform designs – easily adapted by extension of the wings or the addition of another storey – appear to have proved the most popular amongst Boards of Guardians throughout England and Wales.[10]

Critics of the new Poor Law were quick to exploit the apparent similarities between the architecture of the new workhouses and model prisons. The connection was far from an unfortunate coincidence, they argued, for the new designs signalled that poverty was to be treated as a crime. This charge was relentlessly pursued in Wythen Baxter's famous *Book of the bastiles*, a compendium of complaints against the new workhouse system published in 1841.[11] It was also given visual expression in the images published in the second edition of Pugin's *Contrasts* (1841). Here, Pugin portrayed the 'modern poorhouse' as a prison-like Panopticon in which the incarcerated pauper was reduced to a pathetic and powerless figure, robbed of all dignity during life and after death. The 'antient poor hoyse', in contrast, was represented as a spiritual community, a place of sanctuary for the poor and the weak (Figure 4.2).[12] Such iconography clearly had a critical rather than a strictly historical intent; it was supposed to represent, in landscape form, what Pugin claimed to be the degeneration of English moral and aesthetic values. Some features of this romantic, conservative critique were mirrored in contemporary radical attacks on the new workhouse system. A Chartist hand-bill which found its way into Somerset House during 1839, for example, included a graphic sketch of a pauper couple abandoned to their fate outside an anonymous 'Union' (Figure 4.3).[13] In this case, as in Pugin's, the contrast between the imposing outline of the buildings and the frailty of the paupers themselves provided a powerful symbol of the utilitarian calculus said to lie at the heart of the new system. Interestingly, however, the Chartist version explicitly contrasts this 'present' not with a 'past' but with a 'future', in which the imagery of unsympathetic, bureaucratic governance is replaced by that of an educative, patriarchal domesticity.

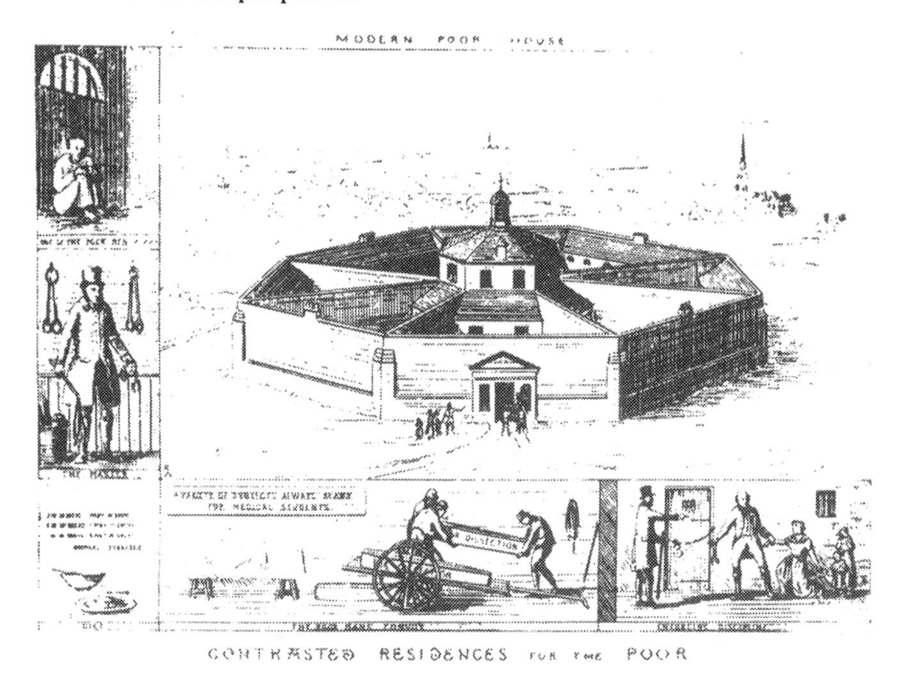

Figure 4.2 Ancient and modern: Pugin's view of the workhouse (A. W. N. Pugin, *Contrasts*, 1841)

Figure 4.3 The present and the future: A Chartist vision (*Source*: C. Mott to J. Lefevre, 13 April 1839, MH 12/15065)

The critics of the new Poor Law thus had little doubt of the deterrent and disciplinary functions of the workhouse system. The Commissioners responded to their attacks by denying that physical punishment played any part in official workhouse policy. It was the need for order rather than punishment, they insisted, which dictated the tone of the new workhouse regulations. Such arguments, ironically, echoed those of the prison reformers of the day; it was discipline rather than punishment that held their attention. Likewise, the Poor Law authorities represented the workhouse as a place where discipline reigned, rather than the caprice of particular individuals; an institution whose inmates and officials were impressed, above all, with the power of the rule. The 1834 Report had famously caricatured the parochial poorhouse as

> a large almshouse, in which the young are trained in idleness, ignorance and vice; the able-bodied maintained in sluggish sensual indolence; [and] the aged and more respectable exposed to all the misery that is incident to dwelling in such a society, without government or classification.[14]

The new Union workhouse, in contrast, was designed to be a disciplinary institution, its inmates subject to the rule of official regulations. It was this disciplinary aspect of central policy which (in the Commissioners' own words) 'led those who opposed the law to stigmatize the workhouse as a bastile'.[15]

Kempthorne's workhouse plans signalled another of the strategies which was to occupy a privileged place in the discourse of workhouse policy after 1834: the strategy of classification. In theory, 'classification' referred simply to the division of the pauper host into meaningful categories, for the purposes of institutional policy; the mapping of what Bentham had once referred to as 'Pauper-land'.[16] In practice, as with Bentham, classification was often identified with a strategy of spatial segregation. The Commissioners' workhouse classification policy required, at the very least, a strict partitioning of male and female wards; in Kempthorne's model plans, the accommodation for these classes was kept entirely separate.[17] Although this policy was pursued in all welfare institutions (in one bizarre instance, there was even a demand for the segregation of corpses in an asylum mortuary[18]), the policy of separating the sexes proved to be particularly controversial in the case of workhouses, as it threatened to divide married couples who might enter the workhouse, at the same time.[19] In 1842, the Commissioners' Workhouse Rules Order specified a minimum classification of workhouse inmates into seven classes: aged and infirm men, able-bodied men over fifteen years of age, boys between seven and fifteen, aged and infirm women, able-bodied women over fifteen, girls between seven and fifteen, and children under seven. The Order stated that 'a ward or separate building and yard' should be assigned to each category of pauper, 'without communication with those of any other class' (cf. Figure 4.1).[20] This was a minimum classification; in practice, the central authority was also to urge on local Guardians the need to segregate sick and fever cases, lying-in women, vagrants and (in later years) the chronically insane. And their statistical returns required the Guardians to make still more detailed sub-divisions of the workhouse population.

The spatial separation of workhouse populations was intended to function in at least three ways: as a basis for appropriate treatment; as a deterrent to pauperism; and as a barrier against contagion, moral as well as physical. The first, the need for special treatment, was perhaps the least emphasised in 1834, although it was later to occupy a more important place in the discourse of Poor Law policy, as we shall see. Critics of official policy were subsequently to claim that the policy of appropriate treatment, requiring the separation of classes, was all but submerged within the strategy of concentration, dominated by considerations of economy and efficiency.[21] While the Commissioners did acknowledge the need for the modulation of treatment according to the needs of each class, the 'deterrent' functions of classification initially received more emphasis. In their early pronouncements, officials insisted that a classified workhouse was also a deterrent workhouse, because the prospect of a segregated and disciplined existence would prevent all save the really destitute from applying for relief.[22] Thus was the logic of the 'semio-technique'[23] of moral regulation translated into workhouse policy. Yet it was the third rationale for classification – as a barrier against 'contagion' – which provided the most enduring justification for the spatial separation of inmates. The moral geometry of workhouse design remained a constant focus of attention throughout the period covered by this book. It was particularly evident in debates over the association of workhouse children with adults, as we shall see in chapter 6. It was the orphan child who was often assumed to be the most vulnerable to 'moral contagion', and it was the prostitute who was generally considered the most 'tainted'. In 1840, the Commissioners even contemplated establishing entirely separate institutions for 'disorderly and lewd women' in London.[24] The failure of this scheme did not, however, signify any diminution in official concern over the demoralising effects of workhouse associations. In 1867, for example, the senior Medical Officer at the Poor Law Board called for the isolation of paupers suffering from venereal disease, 'not so much for the treatment of their disease, as for the separation of persons who carry evidences of profligacy with them, and who are unfit to mix with ordinary cases'.[25]

The strategies of workhouse policy advocated in the early years of the new Poor Law – deterrence, discipline and classification – were not entirely without precedent; indeed, the inculcation of orderly and industrious habits had been an important objective of schemes for workhouse provision ever since institutions of that name had existed.[26] The novelty of workhouse strategies under the new Poor Law lay in the scale of the institutions and in the manner of their regulation; in other words, these were *Union* institutions operating within a workhouse *system*. The formation of Unions permitted larger and more classified workhouses than had generally been possible under the old Poor Law. In the 1770s, the typical number of paupers in each of the two thousand workhouses in England and Wales lay between twenty and fifty, and total workhouse capacity was less than one hundred thousand. In the mid 1850s, however, a reduced number of workhouses catered for a far larger number of paupers (around two hundred and fifty to three hundred inmates each), the total capacity now being over two hundred thousand.[27] Whereas it was quite possible for inhabitants of smaller, more localised workhouses to be familiar figures within their own parishes,[28] those incarcerated in the 'Union' were cut off from the rhythms of everyday life beyond the walls of a distinctive and imposing institution. The transition from parish to Union thus signalled a transformation in the scale and character of workhouse life.

## Models of classification: from concentration to disaggregation?

The concept of classification loomed large in official literature on workhouse management and design after 1834. As we have seen, the term served a number of purposes, even in 1834, and its multiple meanings were to feature prominently in subsequent debates. During the first three decades of the new Poor Law, an increasing amount of official attention was devoted to the treatment of particular classes within 'Pauper-land' – the sick, the insane, vagrants, the young and the old. Eventually, during the 1860s, there emerged a new model of classification, requiring the spatial segregation of pauper classes in detached blocks rather than their concentration in a single workhouse. Such shifts in the interpretation of classification cannot be properly understood in isolation from a variety of wider intellectual, administrative and social contexts. At the local level, the history of workhouse administration was punctuated by the kind of scandals so beloved of the Victorian press. These stories, a few of which are discussed in chapter 9, almost invariably raised the spectre of unregulated association between different classes of pauper in the workhouse – men with women, adults with children, the sane with the insane, the sick with the healthy. They focussed, in other words, on the moral geography of workhouse life. At the national level, the issue of classification provided an important point of departure for conflicts between a variety of organisations and authorities claiming expertise in particular areas of workhouse administration. For the contemporary concern over classification was far from the preserve of the central Poor Law authority alone. Once again, then, a rounded perspective on the history of workhouse policy requires us to look beyond the Poor Law itself in order to locate the 'English bastile' on the landscape of Victorian social policy.

Whereas the central Poor Law authority interpreted classification as a spatial strategy to be developed within the confines of the Poor Law, its critics increasingly used the concept to challenge the very *raison d'être* of the workhouse. On the one hand, it was argued (especially after 1850), the official policy of concentrating all classes of pauper in one workhouse simply failed to discriminate between their different needs; on the other, the bureaucratic ethos of Poor Law administration was excluding a more professional and specialist approach to the treatment of individual classes, such as the young, the sick and the insane. In *Pauperism and Poor Laws* (1852), for example, Robert Pashley attributed the worst defects of the workhouse – which he described as a 'disgrace peculiar to England' – to an absence of effective classification.[29] Writing over a decade later, in very similar terms, the indefatigable Frances Power Cobbe described the workhouse as 'a huge *omnium gatherum* of human want, vice, folly and disease', in which inmates were herded together, irrespective of their particular needs.[30] Cobbe, like many of her associates in the Social Science Association (founded in 1857), interpreted official Poor Law strategy as one dominated by short-sighted deterrence, a policy that was in any case unsuited to the needs of most of the indoor poor. Such critics complained that Poor Law officials, obsessed with large-scale administrative solutions, had effectively ruled out the possibility of more individualised forms of treatment. The workhouse (widely represented as a 'bastile', a 'warehouse' or a 'barracks') was thus portrayed as the child of an unenlightened, bureaucratic approach to social policy, unable to adapt to the lessons of the new social science and inferior in every way to more specialist institutions. According to a writer in the *Social Science Review* in 1865, for example,

> It has none of their best qualities, all of their defects, and so much more defects of its own that it were difficult to enumerate them. As a building, it is a prison without the accommodation of a prison; an asylum without the comforts of an asylum; and a sick hospital with the special proviso that it has no proper convenience for sick people.[31]

The critique of workhouse policy which developed in the halls of social science during the 1850s and 1860s implicated widely-held assumptions (frequently expressed in quite abstract terms) about the relationship between the spatial organisation of institutions and the kinds of regime that could exist within them. These assumptions to some extent rehearsed conventional ideas about the importance of design for strategies of moral regulation (as discussed in chapter 1); but they rejected received wisdom on the precise form institutions were to take. The assault on the workhouse depended on a broader diagnosis of the apparent failure of large 'barrack' designs (in prisons, reformatories, schools, asylums and workhouses) to secure the ultimate aim of 'individualisation'. It was necessary, as one critic of pauper schools put it, to 'break the mass into fragments'.[32] Many of these advocates of smaller-scale institutions subscribed to the so-called 'family principle', which required the disaggregation of larger institutions into 'family'-sized units, either concentrated on one site (in the form of a village or colony), or dispersed throughout ordinary communities. In the words of one advocate of 'cottage' institutions, in 1855, 'If you want show, follow the old plan of association. If you want moral success, follow the new plan of family division'.[33] Although the origins of the 'family principle' lay within the reformatory movement, the campaign for the breaking up of reformatories was extended to institutions in general; 'the practical tendency of the age', argued one reformer in 1871, was 'to diffuse, not to mass, the sick and dependent'.[34] Particular model institutions (such as the reformatory colony at Mettray or the Gheel colony for the chronic insane) were widely promoted as exemplars of the family system, suitable for use in a variety of different contexts in England.[35] In the context of Poor Law policy, it is clear that the 'family principle' played a particularly important role in changing attitudes towards the treatment of pauper children (see below, pp. 97–105).

One of the most significant features of the campaign for the reform of large institutions was the role it gave to women in the design and execution of social policy. The Social Science Association sponsored a number of organisations (including the Workhouse Visiting Society and the Ladies Sanitary Association) which promoted the extension of 'womanly influence' in the running of welfare institutions. Campaigners for the admission of middle-class women into the workhouse, onto Boards of Guardians and, eventually, into the Poor Law inspectorate itself, placed particular emphasis on the virtues of 'domestic economy'. In the sympathetic words of one historian of workhouse policy, these women 'tried to domesticate the system, make it workable, reduce it to human scale, and see its inmates as an extended household. They sought to deconstruct its bricks and mortar into smaller, more homely, and appropriate units'.[36] Many of the women involved in Poor Law debates during the 1860s emphasised the connection between the exclusion of women from workhouse management and the official preference for large-scale institutional solutions. Josephine Butler, for example, argued in her book on *Woman's work and woman's culture* (1869) that

> The tendency at present is to centralization of rule, to vast combinations, larger institutions, and uniformity of system ... For the correction of the extreme tendencies of this reaction, I believe that nothing whatever will avail but the large infusion of Home elements into Workhouses, Hospitals, Schools, Orphanages, Lunatic Asylums, Reformatories and even Prisons.[37]

Four years later, the Local Government Board appointed its very first female inspector, Jane Senior. Her arrival was to prove a landmark in the development of official policy towards workhouse children (see chapter 6).

The 1860s thus mark a period of particularly intense debate about institutional policy in general, and the workhouse in particular. The Social Science Association and its satellites provided vehicles for the promotion of alternative systems and models. The central Poor Law authority, once in the vanguard of the movement for workhouse reform, found itself increasingly cast as an obstacle to further progress. In this context, local disputes could quickly erupt into national controversies, as parochial issues were interpreted in terms dictated by far-reaching debates over the future of the workhouse system. This growing sense of crisis helps to illuminate the causes and consequences of the 'epidemic of scandalous revelations'[38] which was set off by the deaths of several paupers in two London workhouses in 1864–5. In the course of this rapidly developing outbreak of public concern, a new professional body, the Association for Improving the Treatment of the Sick Poor in Workhouses, began its campaign for the reform of Poor Law medical provision. *The Lancet* joined the fray with a highly critical report on the state of London's workhouse infirmaries: 'They are closed against observation, they pay no heed to public opinion; they pay no toll to science. They contravene the rules of hygiene; they are under the government of men profoundly ignorant of hospital rules'.[39] In these years, expertise and officialdom found themselves poles apart; the task of the Poor Law authorities was somehow to reconcile them.

The controversies of the mid 1860s demanded a response from the Poor Law Board. Dr Edward Smith, the Board's principal Medical Officer, was commissioned to carry out a series of intensive inquiries into the condition of workhouses in London and the provinces. The conclusions of his reports mark an important shift in official policy. Many workhouses, it was now accepted, had effectively become 'asylums and infirmaries' by default; 'and so generally is this appreciated that the very term "workhouse" has fallen into disuse, and the word "Union" has been familiarly substituted for it'.[40] The principle of 'less eligibility' was no longer a sufficient basis for Poor Law policy, it was argued, because the vast majority of workhouse inmates were aged, infirm, sick or very young. And what was diplomatically referred to as the 'rapidly advancing views of the age' demanded changes in the design and management of workhouses, particularly in the area of medical provision. The solution, argued Smith, was to provide not one general workhouse but a number of separate buildings, in which each of the various classes of pauper could be housed. In the best modern workhouses, he observed, 'There is no longer one main building in which all the inmates, except infectious cases, may be placed, but a whole village is now comprised within their boundary walls'.[41]

An important consequence of these Reports was that the policy of spatial segregation – the preference for a 'village' of facilities instead of one monster institution – became the official solution to the problem of classification. In 1867, the President of the Poor Law Board sponsored a radical reform of Poor Law provision in London, declaring that the 'treatment of the sick in the infirmaries [must be] conducted on an entirely separate system'.[42] The Metropolitan Poor Act enabled the creation of new Sick Asylum Districts, composed of combinations of the London Unions, in order to allow the establishment of separate infirmaries and a large number of dispensaries. The further creation of a Metropolitan Asylums Board effectively extended the geographical unit of institutional administration from the individual Unions to London as a whole, providing a sound financial and administrative basis, and a larger catchment area, for the construction of large, classified institutions. The Board was responsible for the construction of six isolation hospitals for fever and smallpox patients, and two huge asylums for the chronic insane. Outside London, however, Guardians remained reluctant to combine with their neighbours in the formation of the new district authorities. As a senior central Inspector remarked in 1909, 'There is nothing on which the average Guardian is more sensitive than his local area. He looks upon it as his right, and does not like to be mixed up with his neighbours at all'.[43]

It is significant that the immediate official response to the challenges of the mid 1860s was focussed on London. This provides a marked contrast with the situation in 1834, when it had been the 'Speenhamland' counties that had so obsessed the reformers; at that time, indeed, the government had apparently considered excluding London from the new system altogether.[44] In the 1860s, London led the way, in both the formulation and implementation of workhouse policy. Yet, as the analysis in chapter 5 confirms, the new trends in provision that are apparent in London from the mid 1860s are also evident elsewhere in the country. The central Poor Law authority itself clearly intended that new principles of workhouse management would inform the design of institutional building throughout the country. In 1868, an official circular on workhouse construction issued to every Union in England and Wales recommended that separate day- and night-time accommodation be provided for the sick, the aged, the able-bodied and children, with males and females strictly segregated. (The isolation of prostitutes on moral grounds was once again recommended.) The circular contained specific guidance on a range of matters concerning workhouse design, including ventilation, sanitation, the thickness of the walls and the dimensions of beds, wards and corridors.[45] Such recommendations took on more force with the strengthening of the central architectural inspectorate during the 1870s.

The timing of this new phase of workhouse policy and provision reflects the impact of administrative reforms during the 1860s, as well as debates over policy. Recent research suggests that legislative changes in the financial structure of Union administration effectively enabled the development of less parochial institutional strategies, making local Guardians more receptive to a programme of renewed building activity. After 1861, the basis of parochial contributions to the Common Fund, from which capital expenditure and officers' salaries were financed, was altered from past relief expenditure to present rateable value. The Union Chargeability Act of 1865 transferred the entire cost of out-door and indoor relief to the Common Fund. Historians have argued that these reforms had the effect of stimulating workhouse expenditure, by increasing the significance of the Common Fund, easing financial constraints on Union expenditure and encouraging increased participation on local Boards by more influential sections of the local community.[46] Such changes certainly enabled a renewed phase of workhouse building from the late 1860s, as local studies confirm;[47] yet they did not themselves determine the *form* of the new construction that took place. The following chapter suggests that the increased workhouse expenditure of the 1870s was qualitatively as well as quantitatively distinct from that which preceded it. This difference, clearly evident in the increasing number of specialist institutions being built for particular groups of pauper, reflected the impact of contemporary debates over workhouse policy, which was accelerated by the much-proclaimed crisis of provision in London during the mid 1860s. New policies in fact reinforced the drift of administrative change, for it was increasingly argued (at the regional Poor Law Conferences of the 1870s and 1880s, for example) that the efficient management of specialist institutions required larger areas of administration.[48] The reform of institutional space and the geography of government were thus as closely associated in 1884 as they had been in 1834.

## Conclusion: governing pauperism

The Union workhouse of the late 1830s was designed to accomplish nothing less than a revolution in the moral arithmetic of pauperism. Through its forbidding appearance and its disciplinary regime, the new institution was intended to impress on the poor the virtues of 'independent' labour. It has recently been suggested that the strategy of 1834 was merely one of blind repression, in contrast to the more complex mixture of negative deterrence and positive treatment which characterised policy after 1870.[49] While this argument does capture something of the shift in workhouse policy traced here, it underestimates the extent to which the new Poor Law was conceived as having an educative function from the start. The 1834 reform combined what Foucault calls the 'semio-techniques' of moral instruction with the 'disciplinary' techniques of institutional treatment; while the former predominated in the early years of the new Poor Law, its methods were never simply those of blind repression. 'Throughout the Victorian era', Wiener remarks, 'the Poor Law was "educational machinery", expected to shape popular behaviour. This instructive purpose reached a culmination in the later 1860s and 1870s'.[50] This is not to suggest, as Wiener perhaps implies, that there was a consensus over Poor Law methods during this period. The imposing scale and bureaucratic associations of the workhouse drew increasing criticism after 1850. For the apostles of mid-Victorian social science, the institution was a monument to officialdom and mere containment, rather than a place of expertise and cure. For the advocates of voluntary philanthropy, its monolithic machinery all too often failed to discriminate between the needs of the different classes of pauperism. And for its unwilling working-class clientele, the 'Union' was all the more degrading for its association with rules and regulations. The identification of the workhouse with a bureaucratic system of unprecedented scope was indeed one of the most enduring legacies of 1834.

The history of workhouse policy after 1834 is closely associated with mutations in the official interpretation of the concept of classification. As we have seen, classification provided an important reference point for critics of official policy who, like Frances Power Cobbe, portrayed the workhouse as 'a common *malbolge*' for all the various classes of pauper.[51] The central Poor Law authority responded to increasingly vociferous criticism during the 1860s by re-defining the concept of classification as a technique of spatial segregation. The workhouse was no longer to be a single institution, divided into wards; it was now designed as a 'village' of detached blocks, including cottage homes for children, cells for vagrants, pavilion hospitals for the sick and separate wards for the insane. Although this shift from concentration to disaggregation did not require the abandonment of deterrence as a goal of Poor Law policy, it did allow its most harsh elements to be concentrated on the able-bodied poor. The adoption of more specialist policies towards children, the sick and the insane, was thus to be accompanied by an intensification of the campaign against able-bodied pauperism, especially vagrants.[52] Within the workhouse and its associated institutions, segregation was the key; outside the workhouse, a strategy encouraging greater discrimination between the 'deserving' and the 'undeserving' was being enforced with equal vigour. Classification thus occupied a privileged place in the discourse of workhouse policy. Its re-invention during the 1860s marks one of the most significant moments in the history of nineteenth-century social policy.

# 5

# Building the workhouse system, 1834–1884

The 1834 Act created the conditions for a revolution in workhouse provision. The new machinery of local administration and central regulation was designed to enable the transformation of an inchoate network of Poor Law institutions into an integrated national system. In practice, however, the new institutional landscape was to be far from monolithic; beneath its bureaucratic surface, there continued to be wide variations in the character of workhouse provision. The purpose of this chapter is to outline the historical geography of institutional provision under the new Poor Law during the fifty years after 1834. It is based primarily on an analysis of data contained within a series of unpublished Registers of Authorised Workhouse Expenditure (RAWE), held in the Public Records Office. The RAWE provide a continuous record of officially-sanctioned workhouse expenditure throughout England and Wales, Union by Union. Remarkably, this record has so far escaped the attention of Poor Law historians. And yet it clearly allows an analysis of patterns of workhouse provision which combines a much-needed 'national' perspective with a sensitivity to geographical variations. This analysis overcomes the present dualism in Poor Law history between 'local' studies of implementation (which frequently fail to confront national dimensions of policy change) and 'national' analyses of aggregate data for England and Wales as a whole (which tend to obscure significant variations in patterns of institutional provision). In sum, it puts into proper perspective what one historian has called 'the historical landscape of the national Poor Law'.[1]

## The analysis of workhouse expenditure: problems of interpretation

The importance of the distinction between policy and practice has been one of the guiding principles of much recent Poor Law history. The vast majority of existing studies of Poor Law implementation stress the distance which separated local administrators from the policy-makers in London. Implicit within this approach is the assumption that a national picture of the Poor Law depends primarily on the multiplication of local studies. There have thus been very few attempts to map the spatial and temporal dimensions of the workhouse system at a *national* level; that is, to analyse its historical geography. The prevailing view seems to have been that national-level perspectives inevitably steamroller local variations, obscuring the vast gulf which separated official intention and local reality. This argument, which clearly reflects the widespread reaction against older-style administrative histories, has two main failings. Firstly, it has falsely identified the idea of a national-level perspective with the 'view from the centre', as if national studies must inevitably adopt the perspective of official policy makers. Secondly, as Karel Williams has persuasively argued,[2] it has led merely to a multiplication of local studies, frequently divorced from any direct attempt to consider broader patterns. In this chapter, it is assumed that a national perspective on the workhouse system is not incompatible with a geographically sensitive account of local variations in policy and practice.

The analysis of patterns of workhouse building is not as straightforward a task as it may at first sight appear. The published national-level data, available in appendices to the central Poor Law authority's annual reports, represent merely the capital expenditure authorised to be spent on workhouse 'construction' or (unspecified) 'alterations' in each Union. There is no guarantee, however, that sums authorised were sums actually expended. In some cases, for example, authorisation may not have been followed by actual expenditure; in others, the ultimate local cost of construction may have exceeded the authorised sum. In his analysis of the published data, Karel Williams has shown that such limitations do not altogether rule out an analysis of patterns of workhouse building; indeed, his discussion of expenditure data represents an important advance on previous studies.[3] In order to estimate actual levels of workhouse construction from the published series of authorisations, Williams draws a distinction between Unions in which he considers workhouse construction was 'possible' (those with just a single entry in the annual lists of authorisations) and Unions where construction was 'probable' (those with a series of entries, including authorisations to 'complete' a previously authorised proposal). Williams' analysis is refined by a comparison of the resulting estimates with two Parliamentary returns of existing workhouses recorded in 1839 and 1854.

While Williams' analysis constitutes a marked improvement on previous estimates of workhouse construction, his reliance on published data taken from the appendices of official reports is unnecessarily restrictive. The discussion which follows is based on the unpublished data in the RAWE, which presumably provided the raw material for the annual appendices. The RAWE provide a much more versatile and reliable source for analysis of patterns of workhouse (capital) expenditure. They give occasional details of the locations of individual workhouses, loan arrangements and other incidental details which help us to distinguish those authorisations which were actually followed by construction from those which were not. Alongside certain entries, for example, the words 'rescinded' or 'not acted upon' make it clear where particular authorisations were not followed by construction. In addition, the RAWE describe the type and purpose of every item of authorised expenditure, whereas the published annual appendices indicate merely its quantity. The RAWE thus provide a continuous record of patterns of local expenditure in Unions throughout the country, making it without doubt the most useful source available for the analysis of patterns of workhouse construction.

There are, however, two important qualifications to the analysis which follows. Firstly, the RAWE (like the published returns) record those items of capital expenditure which were officially sanctioned by the central Poor Law authority through the issue of an Order or Instructional Letter. In principle, therefore, it is still possible that this record may misrepresent the *actual* workhouse expenditure of any individual Union. The likelihood of underestimation on a large scale is quite remote, since auditors were empowered to surcharge individual Guardians or other Union officers for any unlawful expenditure from the rates, including workhouse expenditure incurred without central sanction. There are marginal notes in the RAWE indicating that some authorisations were made *post facto*, in order to avoid precisely this outcome. The possibility that the RAWE *over*estimate the actual level of local workhouse expenditure seems more plausible, especially since there is ample evidence of differences between central and local authorities over the implementation of official policy. However, it is clear that of all the Orders issued relating to workhouse expenditure, only those which were legally binding (i.e. signed by two-thirds of the Guardians) found their way into the RAWE; those which failed to gain local consent would simply not have been registered. In addition, as we have seen, marginal notes in the RAWE indicate some cases where legally-binding Orders were issued but for some reason were not actually implemented.

The second qualification concerns the extent to which the RAWE series provides a guide to the *quality* as well as the quantity of workhouse expenditure. One of the potential virtues of the RAWE series lies in the distinctions it makes between general categories of expenditure (construction, alteration, purchase, leasing and renting)[4] and between different types of building (sick wards, vagrant wards, workhouse sites, and so on). This level of detail (generally absent in published statistics) is particularly valuable, because so much contemporary debate on Poor Law matters centred on the classification and separation of workhouse inmates. Unfortunately, however, the RAWE series does not enable us to allocate precise figures to local expenditure on particular types of building. Many of the entries refer to a sum

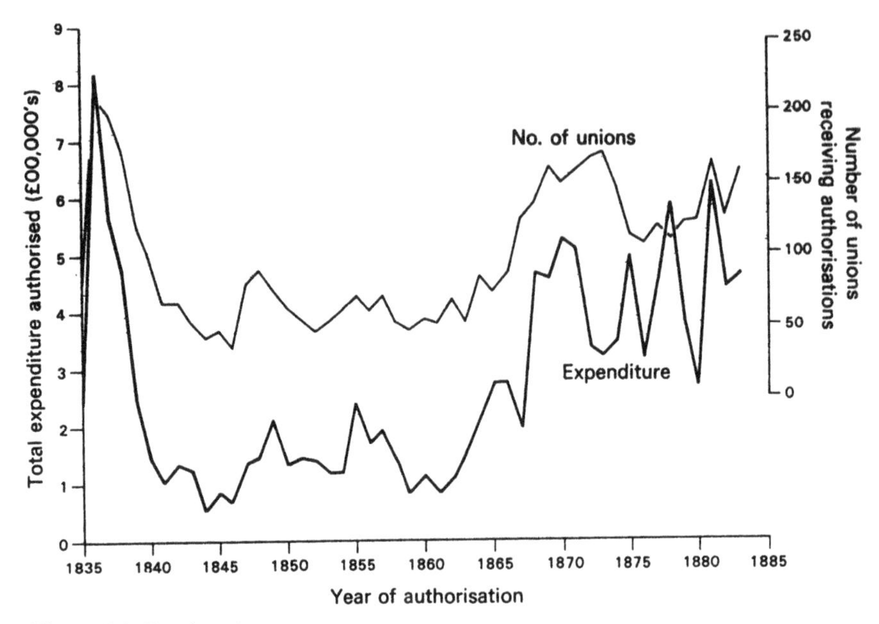


Figure 5.1 Total authorised workhouse-related expenditure, 1835–1883 (*Source*: RAWE)

authorised for several purposes at once without specifying how the total is to be divided (these are described below as *multi-purpose authorisations*). Moreover, general authorisations to alter or build 'a workhouse' may or may not include separate provision for particular classes of pauper; there is no way of telling on the basis of the RAWE alone. Other evidence, in the form of architectural designs for example, suggests that the provision of separate facilities within a newly constructed workhouse complex was more likely after 1850 than before, but it is difficult to give this overall trend a more precise definition. Only where individual RAWE entries refer to the 'construction' or 'erection' of a special ward or building is it reasonable to assume that a new block is in fact separate from the rest of the workhouse.

Nevertheless, the RAWE series constitutes the most comprehensive available record of workhouse-related expenditure under the new Poor Law. For the purposes of the following analysis, each expenditure for which a specific sum is listed in the RAWE was treated as a distinct 'authorisation'. The resulting database of well over six thousand individual authorisations provides a rich source for the study of the historical geography of workhouse provision under the new Poor Law. (See the Appendix to this chapter for a brief explanation of the methods of analysis.)

**Aggregate patterns of workhouse expenditure**

Figure 5.1 shows the aggregate pattern of authorised workhouse-related expenditure in all the Poor Law Unions of England and Wales, estimated from over six thousand individual authorisations for the period between 1835 and 1883 inclusive. The annual figures correspond quite well with published data for authorised workhouse expenditure in the country as a whole,[5] although the latter tend to overestimate actual expenditure (because they include authorisations not actually implemented). The RAWE series shows, broadly, three phases of workhouse expenditure at this 'national' level: a marked peak before 1840, a relatively low (though fluctuating) level between 1840 and 1865, and a significant though uneven increase thereafter. Although these figures are based on current prices, the evidence suggests that adjusting for inflation would only have a marginal effect on these trends (see Appendix). What is very clear is the dramatic increase in expenditure from the mid 1860s. As Table 5.1 indicates, most of this money was devoted to new construction rather than to the alteration of existing buildings, although it should be noted that in some cases 'alteration' could involve a substantial outlay. Many of the authorisations to 'purchase' property do not in fact represent additions to total institutional stock, but merely record the transfer of existing buildings from one Poor Law authority to another. As one would expect, such authorisations appear to have become relatively more important during phases of administrative reorganisation, such as between 1835 and 1839 (the initial period of implementation) and between 1870 and 1874 (following the reform of Poor Law administration in London).

The trend in the number of Unions receiving authorisations (also shown in Figure 5.1) indicates a similar pattern to that of total expenditure, suggesting that peaks in workhouse spending (particularly those of 1835–9 and 1865–74) reflect extensive building activity across a large number of Unions (the majority, in the case of the above two periods), rather than simply a small number of particularly costly projects. Even so, the geography of expenditure underwent a marked shift over the period as a whole. Before 1850, the London Unions accounted for about 11% of total authorised workhouse expenditure; for the period between 1851 and 1866, the proportion was 26%; while between 1867 and 1883 it rose further to 41%. The relative significance of institutional development in London was thus far greater in the 1870s than it had been in the 1830s. Indeed, during three individual years (1870, 1874 and 1883), the London Unions were authorised to spend a greater sum than all the other Unions put together. This pattern reflected the increasing scale and cost of institutional provision in London; elsewhere, though construction was widespread (the most substantial projects being concentrated in the larger cities), it generally involved much smaller sums of money. While regional variations in prices and wage

Table 5.1. *Authorised workhouse expenditure, 1835–1883 (£ 000)*

| | Construction | Alteration | Purchase | Combined | Total |
|---|---|---|---|---|---|
| 1835–39 | 1522 | 246 | 266 | 312 | 2346 |
| 1840–44 | 355 | 138 | 61 | 8 | 562 |
| 1845–49 | 362 | 204 | 79 | 12 | 657 |
| 1850–54 | 388 | 132 | 103 | 36 | 658 |
| 1855–59 | 499 | 157 | 109 | 53 | 819 |
| 1860–64 | 356 | 177 | 90 | 47 | 671 |
| 1865–69 | 987 | 374 | 201 | 116 | 1678 |
| 1870–74 | 971 | 515 | 422 | 132 | 2040 |
| 1875–79 | 1207 | 679 | 195 | 145 | 2226 |
| 1880–83 | 742 | 677 | 290 | 96 | 1805 |
| Total | 7387 | 3301 | 1815 | 957 | 13,461 |

*Note:* Figures are rounded to the nearest thousand pounds. Totals may not correspond due to rounding.

Table 5.2. *Authorised workhouse expenditure, by purpose, 1835–1883 (£ 000)*

| | 1835–50 | 1851–66 | 1867–83 | Total |
|---|---|---|---|---|
| *Single-purpose authorisations* | | | | |
| Construct workhouse | 2154 | 1110 | 1236 | 4500 |
| Alter workhouse | 603 | 453 | 1524 | 2580 |
| Purchase workhouse | 125 | 10 | 95 | 229 |
| Purchase workhouse site | 212 | 171 | 85 | 468 |
| Construct special ward | 154 | 356 | 2101 | 2611 |
| Alter special ward | 21 | 67 | 510 | 598 |
| Purchase special ward | 1 | 10 | 65 | 76 |
| Purchase special ward site | 13 | 54 | 84 | 151 |
| Purchase land | 17 | 71 | 278 | 367 |
| Subtotal | 3301 | 2302 | 5979 | 11,582 |
| *Multi-purpose authorisations* | | | | |
| Purchase site and construct workhouse | 311 | 104 | 19 | 434 |
| Other | 84 | 158 | 1202 | 1444 |
| Subtotal | 395 | 262 | 1221 | 1879 |
| *Total* | 3696 | 2565 | 7200 | 13,461 |

*Notes:* (i) Figures are rounded to the nearest thousand pounds. Totals may not correspond due to rounding.
(ii) 'Multi-purpose' authorisations are those which refer to more than one purpose without specifying how the expenditure was divided.

inflation may have played a part, these figures confirm the increasing scale of institutional provision in London.

Analysis of the RAWE series thus suggests that the period under study was marked by two major phases of institutional building activity, in terms of both the cost of expenditure and the number of Unions engaging in construction; the first was in the late 1830s, and the second beginning in the late 1860s. The RAWE evidence (summarised in Table 5.2) suggests that these two phases were characterised by quite different kinds of workhouse expenditure. The former, a relatively short-lived burst of activity, reflects the initial impact of Poor Law reform, with hundreds of Boards of Guardians replacing existing parish poorhouses with entirely new Union workhouses. The latter, which inaugurated a much more sustained phase of increased expenditure, largely reflects the building of more special wards or separate institutions for particular categories of pauper, including the sick, fever cases, vagrants, children and the insane. In the rest of this chapter, we shall consider in more detail the significance of these patterns of construction with respect to complete workhouses and special wards, respectively.

**Constructing Union workhouses**

Over the entire period between 1835 and 1883, the RAWE series yields a total of 554 authorisations to construct complete workhouses (i.e. institutions catering for all or nearly all classes of pauper). This figure compares well with the best previous estimates of actual construction. Karel Williams, for example, calculates from published evidence that the number of Unions 'probably' building workhouses between 1834 and 1870 lies between 492 and 548;[6] the figure provided by the RAWE series for the same period is 511 Unions (building 521 workhouses). The present analysis, which rests on somewhat firmer evidence, thus supports and refines Williams' general conclusion. The overall pattern of workhouse construction is clear from Figure 5.2: the majority of workhouses (341) were authorised to be built before 1840, with two minor peaks of activity subsequently occurring in the early 1850s (following the administrative and institutional crises of 1847–9) and the late 1860s (following the reorganisation of Poor Law provision in London and elsewhere). As far as complete workhouses are concerned, therefore, it is perhaps inaccurate to speak of more than one 'wave' of construction, as some historians have done;[7] the pattern more closely resembles an immense flood, followed by an irregular drip. The evidence nevertheless confirms Williams' argument that implementation of the official programme of workhouse construction after 1834 was not disabled by the lack of an absolute central power to force workhouse construction on local Boards of Guardians. In the absence of local enthusiasm for a new workhouse, the central authority had a range of weapons at its disposal, including

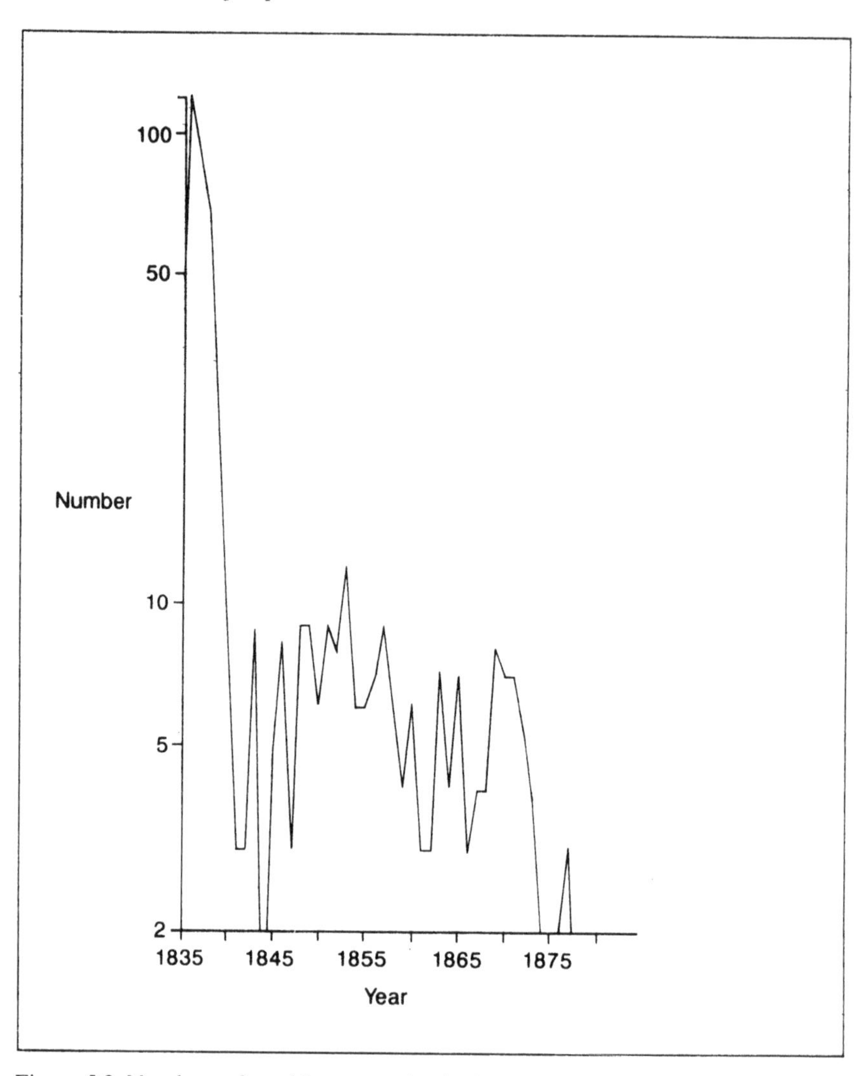


Figure 5.2 Numbers of workhouses authorised to be built, 1835–1883
(*Source*: RAWE)

the threat to order the closure of existing workhouses and the refusal to sanction alternative schemes for improvement of existing facilities. (These tactics were particularly in evidence in a number of Northern Unions during the late 1850s: see chapter 9). During the official campaign for workhouse construction, the Assistant Commissioners took on the role of ambassadors for the central authority. Many of them became adroit strategists, canvassing influential local figures (especially the *ex-officio* Guardians) prior to crucial votes on workhouse expenditure, carefully tailoring their arguments according to time and place. Under pressure such as this, as Assistant Commissioner Charles Mott put it, 'no Union [could] long resist the advantages of an efficient workhouse'.[8]

The pace of workhouse construction during the initial phase of implementation was exceptionally intense. The model plans published in the Commissioners' *First Annual Report* (discussed in chapter 4) provided a guide for some architects; in other cases, local Guardians went their own way.[9] The architect George Gilbert Scott, better known for his more elevated role in the Gothic Revival, is reckoned to have completed about fifty-three of the new workhouses (in partnership with W. B. Moffat) by 1845. Even if these jobs were not especially lucrative, as the architects frequently complained, they were at least plentiful. As Gilbert Scott was later to recall:

> For weeks, I almost lived on horseback, canvassing newly-formed Unions. Then alternated periods of close, hard work in my little office at Carlton Chambers, with coach journeys, chiefly by night, followed by meetings of Guardians, searching out of materials, and hurrying from Union to Union.[10]

The geography of authorised workhouse construction after 1834 was distinctly uneven. Scott's activities, for example, were concentrated in a swathe of southern counties extending from Lincolnshire to Devon. The first map in Figure 5.3, which shows the location of Unions authorised to build workhouses before 1840, confirms the more general picture. Although authorisations were not confined to any one region, the map displays an uneven pattern. It is partly a product of the historical geography of Union formation; the survival of old Poor Law Incorporations, for example, effectively obstructed the building of new workhouses in parts of the West Riding, the Welsh Borders, Norfolk and Sussex (cf. Figure 3.4). The pattern also reflects the resistance of some newly-formed Boards of Guardians, particularly in Lancashire, Yorkshire, Wales and the South West. Figure 5.4, showing the location of twenty-three Unions without a central workhouse in 1847 (excluding the old Poor Law Incorporations), confirms this geography of local recalcitrance; eleven of these were in Wales, five in the industrial North, four in the South West, two in London and one in the North East. Also shown on this Figure are the locations of Unions running more than one workhouse; a considerable proportion of these, particularly in the industrial North, were continuing to use old Poor Law parish workhouses as an alternative to building an entirely new institution (see chapter 9). The Guardians in many of these Unions held out against a new workhouse until the 1860s.

A large majority of Unions in Southern and Eastern England built new workhouses before 1848, many of them in the initial flood of authorisations during the late 1830s.[11] Elsewhere, the triumph of the new system was less

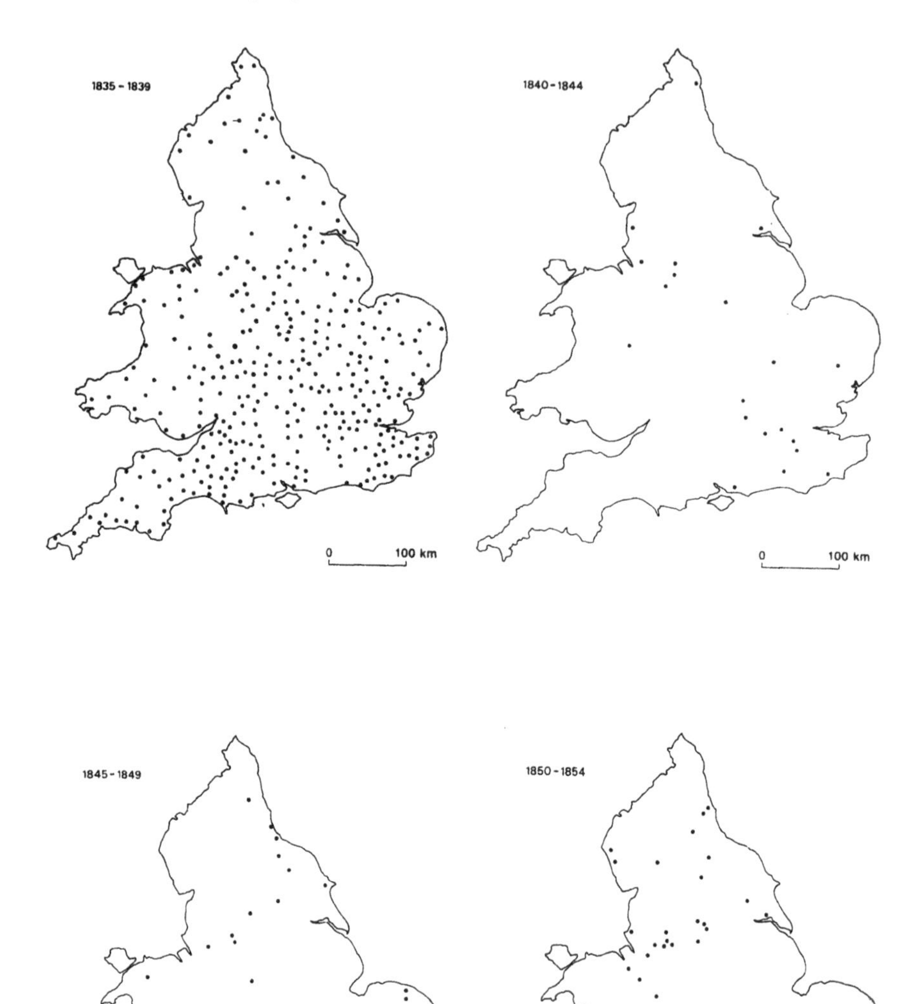


Figure 5.3 The geography of authorised workhouse construction, 1835–1883 (*Source*: RAWE)

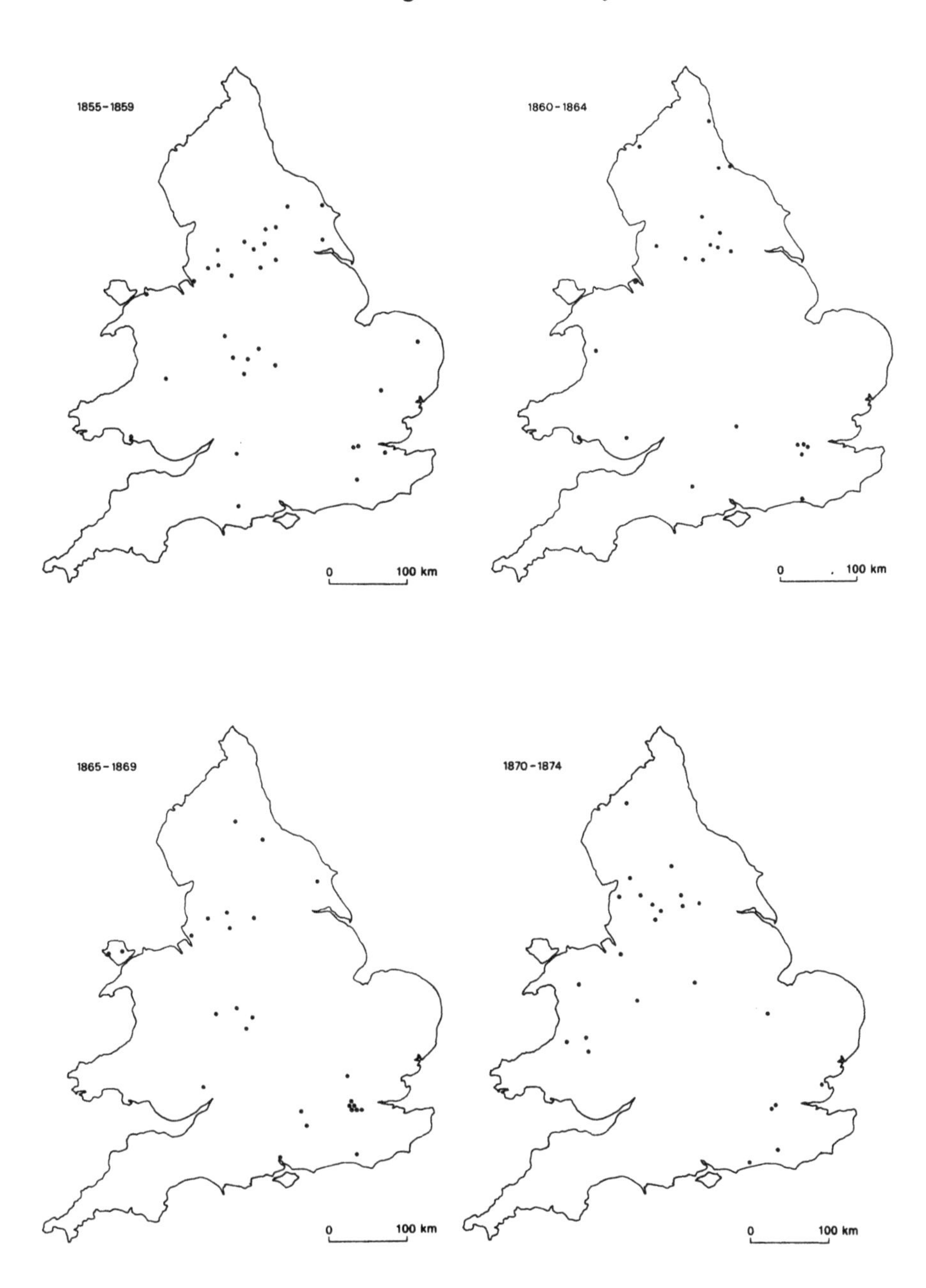
1855-1859
0 100 km
1860-1864
0 100 km
1865-1869
0 100 km
1870-1874
0 100 km

Figure 5.3 (*cont.*)

dramatic. Figure 5.3 graphically illustrates the way in which the 'gaps' in the map of workhouse construction were gradually filled in after the initial phase of implementation. Between 1850 and 1854 (inclusive), for example, authorisations were concentrated in a relatively small number of Unions in industrial Lancashire and the West Riding, the North East, the South West and London itself. At Dewsbury, for example, the Guardians built a new workhouse for the first time in 1853, at a cost of £8,000; a further £2,000 was spent on a separate building for children. In general terms, those Unions authorised to build new workhouses after 1850 fell into one of two categories: those which were entirely newly formed, perhaps after the dissolution of an old Poor Law incorporation; and those which had long held out against official pressure to construct a new workhouse until there was no alternative. Examples of the latter included Rochdale (authorised to build its first new workhouse in 1863) and Todmorden (authorised to build in 1873). In contrast to the period before 1850, when most of the Unions building new workhouses were concentrated in the South and East, those building after 1850 were more likely to be in London, the industrial North, the rural West and Wales. Figure 5.5, which shows the historical geography of workhouse construction in the industrial North, underlines the slow pace of workhouse construction in a substantial number of these Unions (cf. chapter 9). It suggests that there were relatively few new workhouses built in Lancashire and the West Riding before 1848, the majority of Unions remaining without a new workhouse until the 1850s or 1860s. Significantly, those Unions which

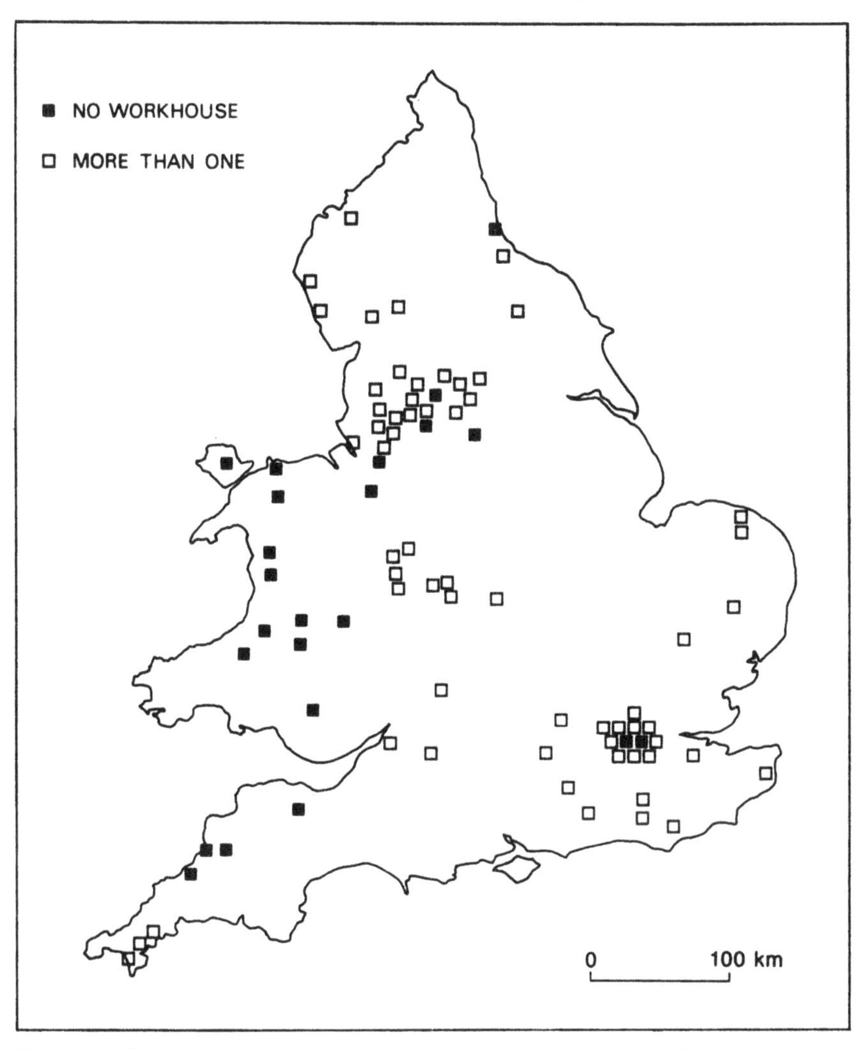


Figure 5.4 Unions with no workhouse, or with more than one, 1847 (*Source*: PLC, *Thirteenth Annual Report*, 1847, Appendix B11)

had been at the centre of anti-Poor Law protest during the 1830s, such as Huddersfield, were amongst the slowest to build new institutions (see chapters 7–9).

**Breaking up the bastile: the trend to separation**

The RAWE series suggests that an increasing proportion of authorised workhouse expenditure was devoted specifically to the construction of wards

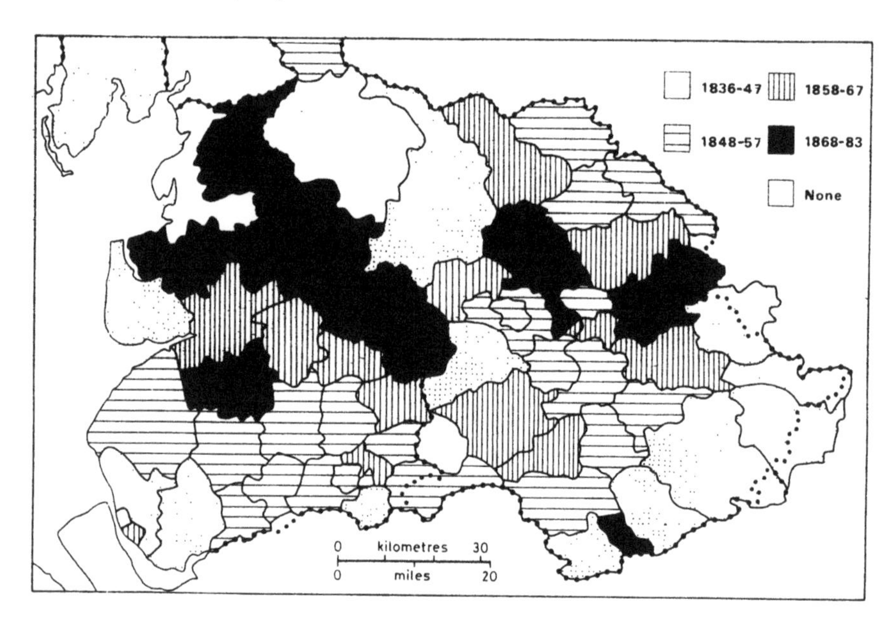


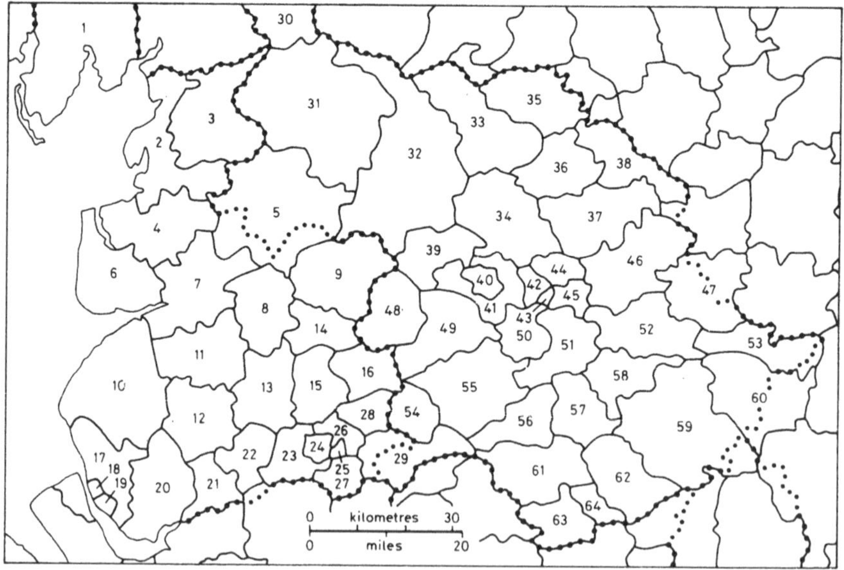


Figure 5.5 Workhouse construction in the industrial North, 1836–1883 (*Source*: RAWE)

Figure 5.5 (cont.)

| *Lancashire* | *West Riding of Yorkshire* | |
|---|---|---|
| 1. Ulverstone | 30. Sedbergh | 59. Doncaster |
| 2. Lancaster | 31. Settle | 60. Thorne |
| 3. Lunesdale | 32. Skipton | 61. Wortley |
| 4. Garstang | 33. Pateley Bridge | 62. Rotherham |
| 5. Clitheroe | 34. Wharfedale | 63. Eccleshall |
| 6. Fylde | 35. Ripon | 64. Sheffield |
| 7. Preston | 36. Knaresborough | |
| 8. Blackburn | 37. Wetherby | |
| 9. Burnley | 38. Great Ouseburn | |
| 10. Ormskirk | 39. Keighley | |
| 11. Chorley | 40. Bradford | |
| 12. Wigan | 41. North Bierley | |
| 13. Bolton | 42. Bramley | |
| 14. Haslingden | 43. Holbeck | |
| 15. Bury | 44. Leeds | |
| 16. Rochdale | 45. Hunslet | |
| 17. West Derby | 46. Tadcaster | |
| 18. Liverpool | 47. Selby | |
| 19. Toxteth Park | 48. Todmorden | |
| 20. Prescot | 49. Halifax | |
| 21. Warrington | 50. Dewsbury | |
| 22. Leigh | 51. Wakefield | |
| 23. Barton | 52. Pontefract | |
| 24. Salford | 53. Goole | |
| 25. Manchester | 54. Saddleworth | |
| 26. Prestwich | 55. Huddersfield | |
| 27. Chorlton | 56. Penistone | |
| 28. Oldham | 57. Barnsley | |
| 29. Ashton | 58. Hemsworth | |

for particular classes of pauper, such as the sick, fever cases, vagrants or children. As Table 5.3 shows, these buildings accounted for the vast majority of Poor Law buildings *specifically* authorised to be built or purchased after 1866. As noted above, this evidence should be treated with considerable caution. The categories of building type listed in Table 5.3 are not mutually exclusive; an authorisation to 'build a workhouse', for example, may or may not have included provision for separate wards. Furthermore, the RAWE series does not appear to include those institutions which were constructed by combinations of Unions, such as District Schools, or Metropolitan Asylums. In addition, 'multi-purpose' authorisations (defined above) account for about 10% of expenditure over the whole period, making it impossible to allocate precise sums to precise purposes in these cases. The latter problem has been overcome by a simple method of estimation, explained in the Appendix to this chapter (see also Table 5.4). Even so, such problems do qualify the usefulness of the RAWE as a guide to trends in the provision of separate buildings for special classes of pauper after 1834.

Bearing these qualifications in mind, the RAWE series identifies some noticeable trends in workhouse provision between 1835 and 1883. Tables 5.2 and 5.3 suggest that after 1866, authorisations involving the construction of separate wards for special classes of pauper became far more significant than

Table 5.3. *Numbers of Poor Law buildings authorised to be constructed or purchased, 1835–1883*

| Building type | 1835–1850 | 1851–1866 | 1867–1883 | Total |
|---|---|---|---|---|
| Workhouse (built) | 402 | 100 | 52 | 554 |
| Workhouse (purchased) | 84 | 20 | 24 | 128 |
| Children's wards or schools | 9 | 25 | 57 | 91 |
| Sick wards* | 56 | 52 | 155 | 263 |
| Fever wards | 21 | 39 | 156 | 216 |
| Insane wards | – | 4 | 14 | 18 |
| Casual or vagrant wards | 21 | 24 | 175 | 220 |
| Receiving or probationary wards | 4 | 2 | 31 | 37 |
| Aged or married couples wards | 1 | – | 5 | 6 |
| Chapel | 4 | 16 | 11 | 31 |
| Relief station | – | 3 | 30 | 33 |
| Offices | 4 | 25 | 20 | 49 |
| Miscellaneous | 28 | 30 | 89 | 147 |
| Total | 634 | 340 | 819 | 1793 |

*Notes:* The figures are taken from all authorisations, including multi-purpose authorisations. They do not include 'follow-up' authorisations to complete a building already authorised to be constructed (see Appendix).
* Includes venereal and lying-in wards.

Table 5.4. *Estimated expenditure on the construction or purchase of special wards, 1835–1883 (£000)*

| | 1835–1850 | 1851–1866 | 1867–1883 | Total |
|---|---|---|---|---|
| Children's wards or schools | 111 | 154 | 503 | 768 |
| Sick wards* | 41 | 118 | 1669 | 1829 |
| Fever wards | 12 | 33 | 162 | 207 |
| Insane wards | – | 8 | 67 | 76 |
| Casual or vagrant wards | 8 | 18 | 208 | 234 |
| Receiving or probationary wards | 1 | – | 26 | 27 |
| Aged or married couples wards | – | – | 3 | 3 |
| Chapel | 1 | 6 | 12 | 20 |
| Relief station | – | 1 | 42 | 43 |
| Offices | 2 | 52 | 81 | 135 |
| Miscellaneous | 26 | 27 | 145 | 198 |
| Total | 202 | 419 | 2919 | 3540 |

*Source:* RAWE
*Note:* These totals incorporate actual figures for single-purpose authorisations and estimated figures for multi-purpose authorisations. The method of estimation is explained in the Appendix. Totals may not correspond due to rounding.
* Includes venereal and lying-in wards.

those for completely new workhouses. The most important of these wards, in financial terms, were the sick wards or workhouse infirmaries, which accounted for more than half of the total authorised expenditure devoted to special wards over the entire period (cf. Table 5.4). Although a substantial number of these were built before 1866, the rate of construction increased dramatically thereafter, with 155 sick wards or infirmaries being authorised between 1867 and 1883. About the same number of fever wards or isolation hospitals for cases of contagious disease were authorised to be built after 1866, although these were in general much smaller in scale. As we saw in chapter 4, the campaign for improvement of medical provision within the Poor Law played a central role in the policy debates of the mid 1860s.[12] Urban Unions, especially those in the Metropolis, were to be particularly active in the provision of infirmaries during this period. In a process greatly accelerated by the creation of joint authorities under the 1867 Metropolitan Poor Relief Act, the London Unions embarked on a large-scale programme of building, in order to improve separate provision for the sick. After 1870, in particular, expenditure on new infirmaries was very substantial.[13] The Lambeth Union, for example, was authorised to spend £50,000 on a new infirmary at Kennington in 1875 (the final cost was nearer £70,000).

While provision for the sick dominated expenditure on special facilities between 1867 and 1883, a large number of other new institutions were constructed, including vagrant wards (175 authorised to be built in 162 Unions), children's wards, homes or schools (57 authorised in 52 Unions), and lunatic wards (14 authorised in 13 Unions). The building of vagrant wards reflected the growing severity of official policy towards 'casuals'; in many Unions, new construction was accompanied by the introduction of more disciplinary regimes for vagrants. A tougher attitude towards vagrancy is evident from the mid 1860s, when central inspectors emphasised the problems posed by what they described as the 'professional vagrant'.[14] The Pauper Inmates Discharge and Regulation Act of 1871 established stricter regulations on the movement of vagrants in and out of the workhouse, and provided explicitly for the provision of separate deterrent wards, which the Local Government Board recommended to be built on the new 'cellular' system.[15] The construction of new vagrant wards proceeded apace, particularly in the larger towns of the industrial North, including Huddersfield and Halifax, for example; by 1904, over two-thirds of all Unions provided separate cells.[16] In many cases, these cells were divided into two; a small cell for sleeping (characteristically a mere four feet wide) and another for working, the most common tasks being stone-breaking and corn-grinding. Conditions in the unsegregated, 'associated wards' were no more pleasant, as many self-proclaimed 'amateur casuals' discovered during the 1880s.[17]

One of the most striking aspects of this trend towards separate provision is

Table 5.5. *The construction of special wards in the London Unions, 1867–1883*

| | Number | Estimated expenditure (£000) | Proportion of national total (%) |
|---|---|---|---|
| Children's wards or schools | 5 | 176 | 35 |
| Sick wards or infirmaries* | 25 | 810 | 49 |
| Fever wards | 6 | 3 | 2 |
| Insane wards | 5 | 11 | 16 |
| Casual or vagrant wards | 16 | 83 | 40 |
| All Special wards | 128 | 1352 | 46 |

*Note:* See also Table 5.4 and Appendix
* Includes venereal and lying-in wards

its geography. Although a large number of Unions were authorised to build separate wards during the period after 1866, the largest and most costly of the new facilities were mainly to be found in the London Unions. As Table 5.5 indicates, 46% of authorised expenditure specifically devoted to the construction of separate buildings between 1867 and 1883 – other than entire workhouses – was accounted for by the London Unions. (As indicated above, some but by no means all of this pattern may reflect regional differentials in building costs.) The significance of Metropolitan expenditure was particularly marked in the case of new sick wards and vagrant wards. The Capital's apparently less dominant role in spending on new facilities for the insane and fever cases may partly be explained by the fact that the RAWE series excludes institutions built by district authorities, including the Metropolitan Asylums Board, established in 1867, responsible for the establishment of several large asylums and fever hospitals. Outside London, the more substantial institutional projects sanctioned after 1867 were generally undertaken in the larger urban Unions, often at a distance from existing workhouse facilities. Large-scale provision for the insane, for example, was concentrated in cities like Manchester, Bradford, Oldham and Portsmouth, as well as in London itself (see below, pp. 109–10).

Although the London Unions account for a large proportion of total expenditure after 1867, building activity on a smaller scale was common in a large number of provincial Unions. Figure 5.6 shows the geography of authorised construction in the case of children's wards and schools, a category (as Table 5.4 shows) which accounts for over 20% of the total expenditure separately authorised for all special wards. Although this is only a sample of the total number of Unions housing pauper children in separate

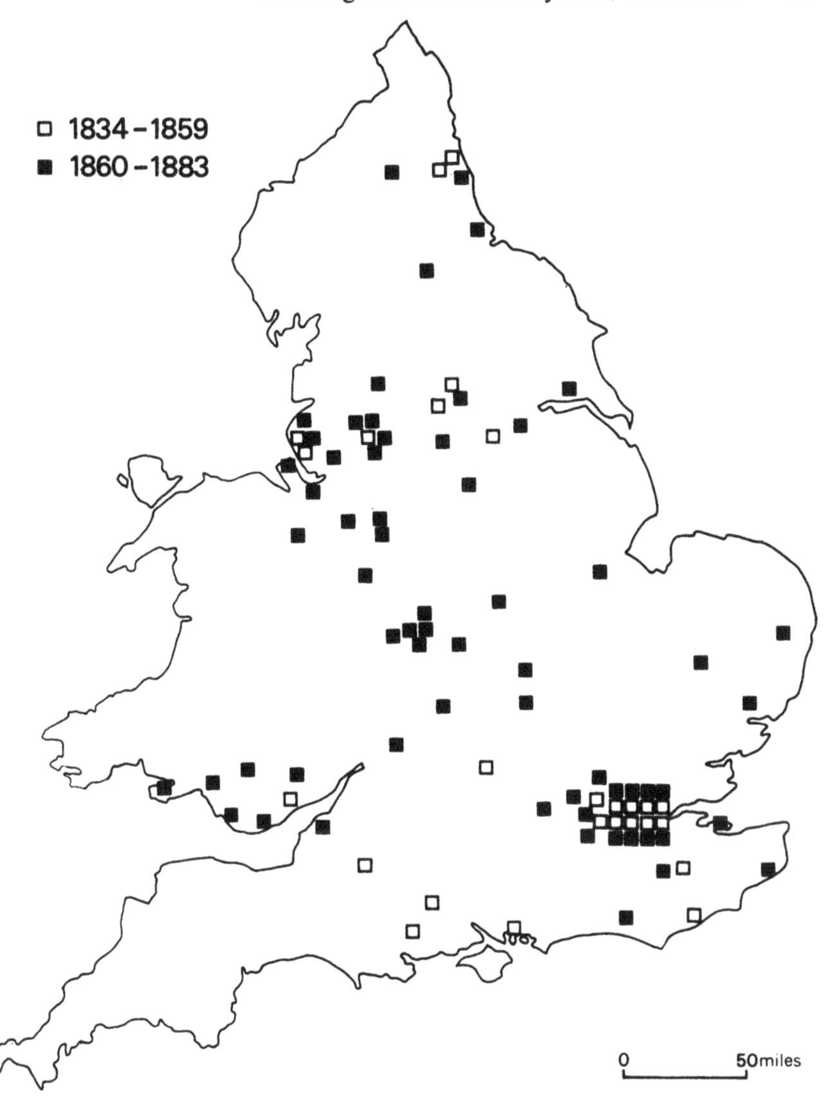


Figure 5.6 Unions authorised to build children's wards, homes or schools, 1834–1883 (*Source*: RAWE)

accommodation (many of which may have converted existing buildings or incorporated children's wards within an authorisation to build a new workhouse), it does demonstrate that London had no monopoly on such facilities. It is equally clear that institutional provision for pauper children took a wide variety of forms at different times and in different places. During the 1840s, for example, the Guardians at Leeds, Liverpool and Manchester invested considerable sums in the building of large separate schools (cf. Figure 6.1). Such large institutions were exceptional, however, and after 1870 the building of large institutions for pauper children was virtually confined to London. Between 1867 and 1883, children's institutions authorised to be built in London required an average expenditure of £35,185 per Union, confirming that these were large-scale projects; elsewhere, however, the figure was only £6,292. After 1870, an increasing number of authorisations involved the construction or conversion of small 'cottage homes', such as those established near Bridgend, in South Wales, during 1877.[18] The reasons for the increasing popularity of this design are discussed in the following chapter.

## Conclusion

The analysis in this chapter highlights a number of important dimensions of the historical geography of the workhouse system after 1834. It confirms, in more precise terms than hitherto available, the size and extent of the building programme launched in the later 1830s. It underlines the geographically uneven impact of local resistance to the construction of new workhouses, which was concentrated in Wales and the industrial North. And it also suggests that the new phase of building activity from the late 1860s marked an important moment of transformation in the workhouse system as a whole, not least because the new level of aggregate expenditure was sustained, with marked annual fluctuations, until the late 1890s.[19] The timing of this second phase of construction must be seen in the context of both administrative reform and policy debates (cf. chapter 4). Changes in the financial structure of Union management, following the chargeability reforms of the early 1860s, as well as the availability of cheaper loans, made local Guardians more receptive to a renewed programme of institutional building.[20] In London, the 1867 Metropolitan Poor Relief Act transformed the financial basis of institutional provision, stimulating a less parochial approach to workhouse expenditure.

Changes in the financial basis of Union administration thus appear to have triggered a second phase of workhouse construction from the late 1860s. However, while they enabled renewed building activity at a local level, these administrative changes did not themselves dictate the *form* in which new provision appeared. The analysis presented here suggests that the increasing level of workhouse expenditure after 1865 (especially in London) simultaneously marks a shift in the interpretation of the strategy of classification, from concentration to disaggregation. As new wards appeared alongside older buildings, the space of the Union workhouse was being broken down and dispersed; the 'bastile' was effectively becoming a 'village'.[21] The new geography of 'pauper-land' reflected the impact of contemporary debates over the Poor Law in general, and classification in particular. The following chapter explores two segments of these intertwining discourses – those which concerned the training of pauper children and the treatment of the insane.

## Appendix: Analysis of authorised workhouse expenditure

The discussion in this chapter is based on the analysis of a series of Registers of Authorised Workhouse Expenditure (RAWE) stored at the Public Records Office, Kew. The entries in the first three volumes of the RAWE series (covering the years between 1835 and 1883 inclusive) were processed and converted into a data base containing a total of 6,629 rows and 4 columns. The rows represent the total number of separately identifiable authorised expenditures (not including leases or hirings) in 658 Unions between 1835 and 1883. The columns are constituted by four variables: a number for each Union, the date of authorisation, the amount of expenditure authorised and a code representing the purpose of the expenditure. As well as identifying each type of building listed in the RAWE ('workhouse', 'sick ward', 'schools', and so on), these codes were designed to distinguish between (1) construction, alteration and purchase; (2) initial authorisation and subsequent 'follow-ups' (i.e. those authorisations completing or adding to an initial authorisation); and (3) single- and 'multi-purpose' authorisations (i.e. those in which one sum of money was authorised for several purposes). Processed in this way, the RAWE constitute a flexible and accessible source for the study of both general trends in authorised capital expenditure and particular features of institutional provision under the new Poor Law.

A relatively simple computer programme enabled analysis of temporal and spatial variations in authorised workhouse provision, at various scales. One question concerned the frequency with which particular types of building were authorised (excluding 'follow-ups') in particular Unions and/or over various periods of time. Another question focussed on trends and patterns in expenditure, for various combinations of building-types, Unions and periods. Apart from the potentially problematic relationship between authorised and actual expenditure, which is discussed in the text, this analysis of expenditure presented two problems. There was, firstly, the problem of the multi-purpose authorisations, which account for about 10% of total expenditure. The method used for estimating the proportion of the expenditure to be allocated to each particular purpose in these cases was based on the assumption that, for any given period of reasonable length (intervals of fifteen years were used here), the average expenditure for the construction of each individual type of building was identical in single-purpose and multi-purpose authorisations. Table 5.4 incorporates the results of such estimates of expenditure authorised within 'multi-purpose' authorisations. While these estimates are subject to error, particularly where the number of buildings concerned is small, their incorporation in the analysis does give a better indication of the significance of special-purpose wards than do single-purpose authorisations alone. The estimates were checked against the total sums recorded in multi-purpose authorisations (which include expenditure not only for construction, but also for alterations and purchases) and were found to be plausible.

The second problem considered in the analysis of workhouse expenditure was the possibility of variations in unit building costs over time and space. The figures presented here are based on current prices only. Available time-series of general building costs, estimated from input prices, suggest that inflation was a relatively insignificant factor over the period as a whole (the falling cost of materials offsetting the rising costs of labour), with the exception of the mid 1870s.[22] An adjustment based on an estimate of the real costs of workhouse construction would have a very marginal effect on patterns of expenditure after 1872, but would not significantly affect the general conclusions advanced here. Regional variations in the cost of building are more difficult to handle, but it is unlikely that these would account for the very marked shift in expenditure towards London after the mid 1860s.

# 6

# Classifying the poor: maps of pauper-land

The concept of classification lay at the heart of many of the debates over social policy during the nineteenth century. In the context of workhouse policy under the new Poor Law, classification raised a potential problem – the mixing up of different 'classes' of pauper – without defining any specific solution. After 1834, there was to be endless debate over the suitability of particular institutional designs and regimes for the various classes of indoor pauper. These debates involved a constant flow of ideas and influence across the boundaries of the Poor Law, as a wide variety of authorities and organisations claimed a role in the treatment or care of workhouse inmates. Some critics argued that it was necessary to redefine these very boundaries, in order to allow a new map of 'pauper-land' to be drawn. Such issues of expertise and authority provided the raw material for many of the Poor Law debates of the period, at Social Science Association congresses, in the professional journals and at official inquiries. In order to explore the sources and character of this discourse of classification in more depth, it is useful to focus on those regions of pauper-land whose boundaries were most in dispute. This chapter thus considers debates over classification in two key areas: the training of pauper children and the treatment of insane paupers. Here were two of the most dependent classes of indoor pauper, truly wards of the state; here too, the problem of pauperism became one of government and classification.

## The contagion of pauperism: the training of pauper children

> The geographical proximity of the children to adult paupers is an objection which has been much exaggerated. Children do not catch pauperism as they catch measles, by passing someone on the stairs.[1]

During the mid-nineteenth century, children under the age of sixteen constituted between one-third and two-fifths of the pauper host.[2] Although only a fraction of them were inmates in Poor Law institutions, it was the care of these indoor children – especially those without parents able or willing to support them – which commanded by far the most official attention. During the fifty years after the 1834 reform, a whole range of different schemes were advocated, from the building of huge training schools for hundreds of inmates, to entirely non-institutional solutions, as in the policy of boarding-out. Debate over the relative merits of these systems reached a peak in the 1860s, a decade in which reformers and administrators seemed to be permanently engaged in statistical warfare over the relative merits of various methods of training. Virtually all the participants in this debate shared two basic assumptions: first, that the design of institutions was a critical influence on the success or failure of pauper training; and, second, that it was necessary to isolate the pauper child from the tainted associations of the workhouse.

The first detailed indication of official policy towards indoor pauper children under the new Poor Law was provided in 1838, with the publication of two special reports by James Kay, an Assistant Poor Law Commissioner. Kay took great care to explain his reasons for advocating special treatment for pauper children; this class above all others, he argued, deserved more than deterrence, since children were pauperised 'not as a consequence of their errors, but of their misfortunes'. The logic of 'less eligibility' could not, therefore, apply directly to them. What they required was a programme of education, which was described as 'one of the most important means of eradicating the germs of pauperism from the rising generation'.[3] This programme was to consist primarily of industrial training and moral discipline; a regime designed to transmit the values of 'independent' labour. Kay intended the industrial training to be less an apprenticeship in a particular trade than an inculcation of 'the practical lesson that they [pauper children] are destined to earn their livelihood by the sweat of their brow'.[4] For boys, this meant manual and agricultural work of various kinds; for girls, household and other domestic duties. The moral discipline was designed to inculcate the virtues of obedience, honesty, labour and punctuality, 'sentiments and habits' (it was claimed) 'foreign to the class to which they belong'.[5] Kay and his colleague, Edward Tufnell, placed great emphasis on the value of discipline in both work and play. They argued that the use of fixed timetables, regimented drills, out-door labour, gymnastics and musical instruction would impart practical moral lessons:

> In giving the child an erect and manly gait, a firm and regular step, precision and rapidity in his movements, promptitude in obedience to commands, and particularly neatness in his apparel and person, we are insensibly laying the foundation of moral habits most intimately connected with the personal comfort and the happiness of the future labourer's family.[6]

Kay argued that such a training was impossible in existing workhouses, where the children could never be completely separated from the adults; even in the Union workhouse, they were bound to be exposed to moral 'contamination'. Tufnell put the point still more forcefully:

> The atmosphere of a workhouse is tainted with vice; no one who regards the future happiness of the children would ever wish them to be educated within its precincts.[7]

Both Kay and Tufnell were soon convinced of the need to establish entirely separate institutions for pauper children. During the 1840s, they advocated the formation of special combinations of Unions which would enable the construction of large district institutions. Legislation in 1844 and 1848 permitted the formation of such district authorities and a number of Unions proceeded to form school districts.[8] Outside London, however, the official campaign for the establishment of district schools was frustrated by the reluctance of individual Boards of Guardians to adopt the provisions of these Acts. The number of district schools was never large. In 1860, for example, there were only six, catering for about one in every thirteen indoor pauper children; the rest were housed either in ordinary workhouses or in one of the nineteen separate schools built by individual Unions, such as the Moral and Industrial School established by the Leeds Guardians in 1848 (Figure 6.1).[9] Nevertheless, the district school policy remained the preferred system of most, if not all, of the official 'experts' on pauper education until at least 1870. The five Inspectors appointed by the Committee of Council on Education in 1847 to supervise pauper schools and their teachers were left in no doubt that it was their task to promote the adoption of the district school system throughout the country.[10]

Outside official circles, the district schools policy proved to be deeply unpopular. Louisa Twining, founder of the Workhouse Visiting Society in 1856 and an advocate of the 'family system' (discussed in chapter 4), argued that the sheer size of the district schools ruled out any possibility of *individual* treatment; as she told the Social Science Association in 1861, 'we have yet to find that hearts are reached by this system'.[11] Another critic likened life within the district schools to 'a piece of music of endless length, from which all accidentals, or changes in time or key, were rigidly excluded'.[12] The dull, unnatural routine of the large district school (it was claimed) failed to provide the industrial training necessary for girls sent out to domestic service; according to Mary Carpenter, for example, many of them emerged 'quite unacquainted with the way to boil a potato or make a common family pudding'.[13] All these problems were commonly attributed to a fundamental design fault in the principle of 'aggregation', embodied in the 'barrack' system. The alternative model was the 'family system', a regime in which the children would be treated individually rather than *en masse*. In the words of Florence Davenport Hill, author of the influential *Children of the state*, 'All experience in the treatment of this class shows that the nearer the individualisation and interdependence of the family can be approached, the

Figure 6.1 Leeds Moral and Industrial Training School, 1848 (*Source*: Leeds District Archives)

greater the probability of reformation'. Division into small 'family' units, she argued, would allow a more effective balance to be struck between the centripetal powers of 'mutual affection and responsibility', on the one hand, and the centrifugal forces of 'individualisation', on the other.[14] What is striking about such arguments is the degree to which both the critics and defenders of official policy conceived the problem of the workhouse child in terms of the power of design, disciplinary training and the reformation of conduct; this was, undoubtedly, a discourse of moral regulation.

As was suggested in chapter 4, the 'family principle' was intended to be applicable to a wide variety of institutional settings. Indeed, prior to its promotion in the field of Poor Law policy, it had been a powerful weapon in the hands of those campaigning for the establishment of juvenile reformatories. Here the paradigm of the family principle was said to be a celebrated model reformatory at Mettray, in France, where the children were divided into separate houses and 'families', rather than being congregated in a single building.[15] Debates over reformatories spilled over into the domain of Poor Law policy, and vice versa. Many of those urging the virtues of the 'family system' on Poor Law officials, including Florence Davenport Hill, Louisa Twining and Mary Carpenter, were also active in the reformatory movement. Armed with new conceptions of delinquency and moral training, the advocates of the 'family principle' portrayed juvenile pauperism and juvenile crime as symptoms of a deeper moral malaise. Thus Mary Carpenter argued in 1852 that there was 'very little difference between the physically destitute child who is thrown on the parish, and the morally destitute one who is placed in the reformatory school'; lacking an effective family discipline, both require state-sponsored moral training to 'make them good citizens'.[16] Carpenter was not alone in drawing parallels between these two classes. In the same year, Joseph Fletcher, the Education Inspector and Secretary to the London Statistical Society, published a paper extolling the virtues of reformatory 'farm-schools' (especially Mettray) on the continent of Europe. He described the system as eminently applicable to *all* classes of dependent children, 'whether pauper, morally endangered or actually delinquent'.[17] Although it would be wrong to assume that such remarks effectively erased all the differences between policy towards juvenile criminals and pauper children, they do suggest that the distinction between the two was frequently blurred. This traffic in ideas was encouraged by the foundation, in 1857, of the National Association for the Promotion of Social Science. It gave many critics of Poor Law policy a platform for the promotion of new strategies of moral training, including the family system. Lessons learnt in the context of reformatory science were frequently held to be applicable in the training of pauper orphans, and vice versa.[18] One of the legislative fruits of these conceptual borrowings was the 1857 Industrial Schools Act, which made provision for the moral training of a wide class of non-criminal children.

The onslaught on the 'barrack' system for criminal and pauper children, launched in the name of the 'family principle', drew a variety of responses in official circles. In general, the prisons inspectorate regarded the Mettray system as a supplement rather than a replacement for existing models of confinement. However, the appointment of a new Inspector of Reformatories following legislation in 1854 brought Sydney Turner, a known admirer of the 'family system', into the heart of the Home Office.[19] Officials responsible for educational policy were divided between those who supported the development of farm-schools on the model of Mettray (such as Joseph Fletcher) and those who believed the system was inherently restricted to juvenile criminals (such as Edward Tufnell). From the early 1850s, these official attitudes were subject to increasing scrutiny as the campaigns for workhouse reform and the reformatory movement gained pace. Having portrayed itself as the harbinger of a new and progressive system of pauper education, the inspectorate of pauper schools found itself under increasing attack. Possibly the most damaging assault emerged from within the education establishment. In 1851, Joseph Fletcher warned his Poor Law colleagues that 'the experience of the world is opposed to large schools . . . , each with its palace of brick and stone, and its comparatively mechanical method'.[20] The pauper schools inspectorate responded to such challenges in two ways. Firstly, following Tufnell's lead, they dismissed proposals to replace the district schools with disaggregated 'cottage' homes as both unnecessary and inefficient. Larger schools were not only more economical, they argued, but they also ensured a more complete isolation of pauper children from the sources of 'moral infection', outside as well as inside institutions.[21] Secondly, they attempted to defuse the impact of the Mettray model (the paradigm of the 'family system') by pointing to similarities between the regime at Mettray and that in the district schools. In both, out-door labour was regarded as an important means of instilling habits of industry in place of idleness;[22] and in both, notwithstanding the claims of many of Mettray's English admirers, the virtues of order and discipline were paramount. The drills, bands and training ships of the district school system all had their parallels at Mettray. Indeed, according to one inspector, Mettray's family system was 'but a subordinate classification [resembling] the division of a regiment into companies, united together by common feeling and common subordination to the field officers'.[23]

The official defence of district schools was weakened by administrative disputes between the Poor Law Board and the workhouse schools inspectors,[24] and undermined by the misgivings of at least one of the inspectors themselves. In the course of his work during the 1850s, inspector Thomas Browne was converted to the view that the 'family principle' was after all applicable to pauper children. This change of heart perhaps had less to do with an endorsement of the Mettray system than with Browne's growing doubts about large district and separate schools such as those at Leeds and Liverpool, which (he argued) were smothering the individuality of the children and paralysing the moral influence of their teachers.[25] These criticisms were reinforced during the 1860s, with increasing revelations of mismanagement, disease and overcrowding at the London district schools. Eventually, in 1873, in the aftermath of a major review of institutional provision in London, the President of the Local Government Board asked Jane Senior to provide a 'woman's view' of the treatment of pauper girls in the London pauper schools. Her report, a landmark in the history of official policy, came out heavily against the larger district schools and in favour of small 'cottage homes', on what she herself called 'the Mettray principle'.[26]

Jane Senior's advocacy of 'cottage homes', alongside other policies such as boarding-out, should be seen in the context of the growing strength of the campaign for the 'family system' during the 1860s and 1870s (also discussed in Chapter 4). Amidst intensifying criticism of its policies towards pauper children (amongst others), the central Poor Law authority was under great pressure to consider alternatives, including the provision of separate 'family' homes for pauper children. In 1873, it despatched one of its most senior inspectors (Andrew Doyle) to Mettray itself, the Mecca of the reformatory movement.[27] The immediate origins of this decision to send an official to Mettray are unclear, although the publication of Hill's *Children of the state* in 1868 and Jane Senior's report in 1873 must have played a role. What is clear is that Doyle's glowing report was regarded as a turning point by campaigners for Mettray-type alternatives to the district school system. 'Your first impression upon entering', he reported, 'is that you are in a well arranged village, amongst villagers, the members of the families engaged at their various occupations, passing to and fro from house or shop, as in ordinary social life'.[28] For Doyle, Mettray offered the prospect of a less regimented, more 'natural' environment for the training of pauper children.

The retreat from the district schools policy was confirmed in 1874 by the resignation of Tufnell, and again in 1878 by the publication of another official report promoting the adoption of 'the home or cottage system of training' for pauper children.[29] The designs displayed in this report contrast markedly with those of the previous generation of district and separate schools (Figure 6.2). Significantly, all the homes visited in the course of this inquiry (including industrial homes, orphanages and reformatories) were, like Mettray, originally designed for non-pauper children. Once again, institutional strategies were diffusing across the borders of 'pauper-land'. The 1878 Report followed its more celebrated predecessors in endorsing the trend towards the building of smaller cottage homes for pauper children. Between 1870 and 1914, nearly two hundred such homes were authorised to be built by local Guardians.[30] The most ambitious were the 'grouped homes', which consisted (as did Mettray) of a colony of detached 'houses' on

**PRINCESS MARY'S VILLAGE HOMES,**

**ADDLESTONE.**

*Site Plan.*

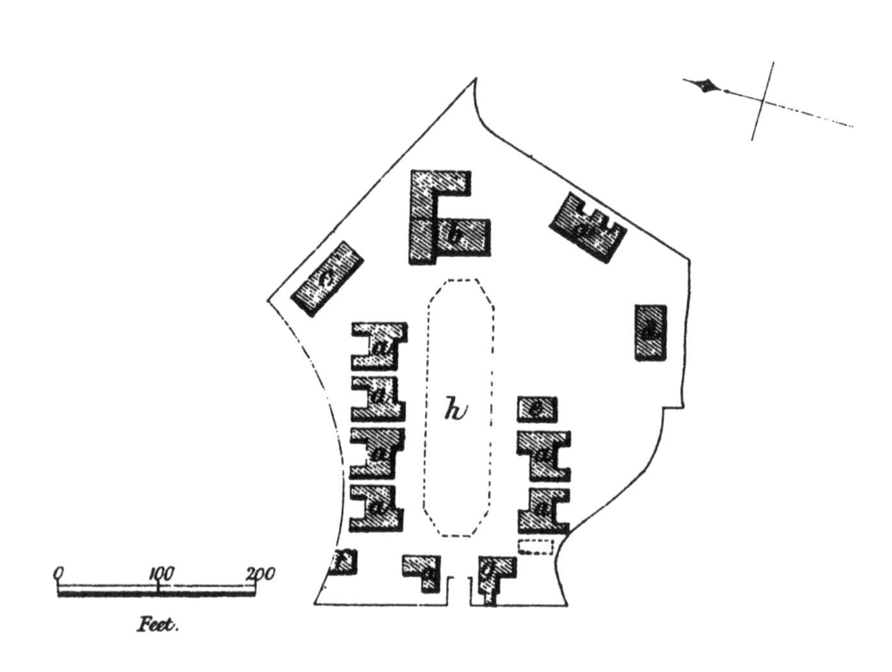


a, a. Girls' Cottages.
b. Schools.
c. Laundry
d. Infirmary.
e. Mission Room.
f. Bath house.
g. Office and Stores.
h. Playground.

Figure 6.2 Cottage homes for orphan children, 1878 (*Source*: Report on Cottage Homes, PP 1878 LX)

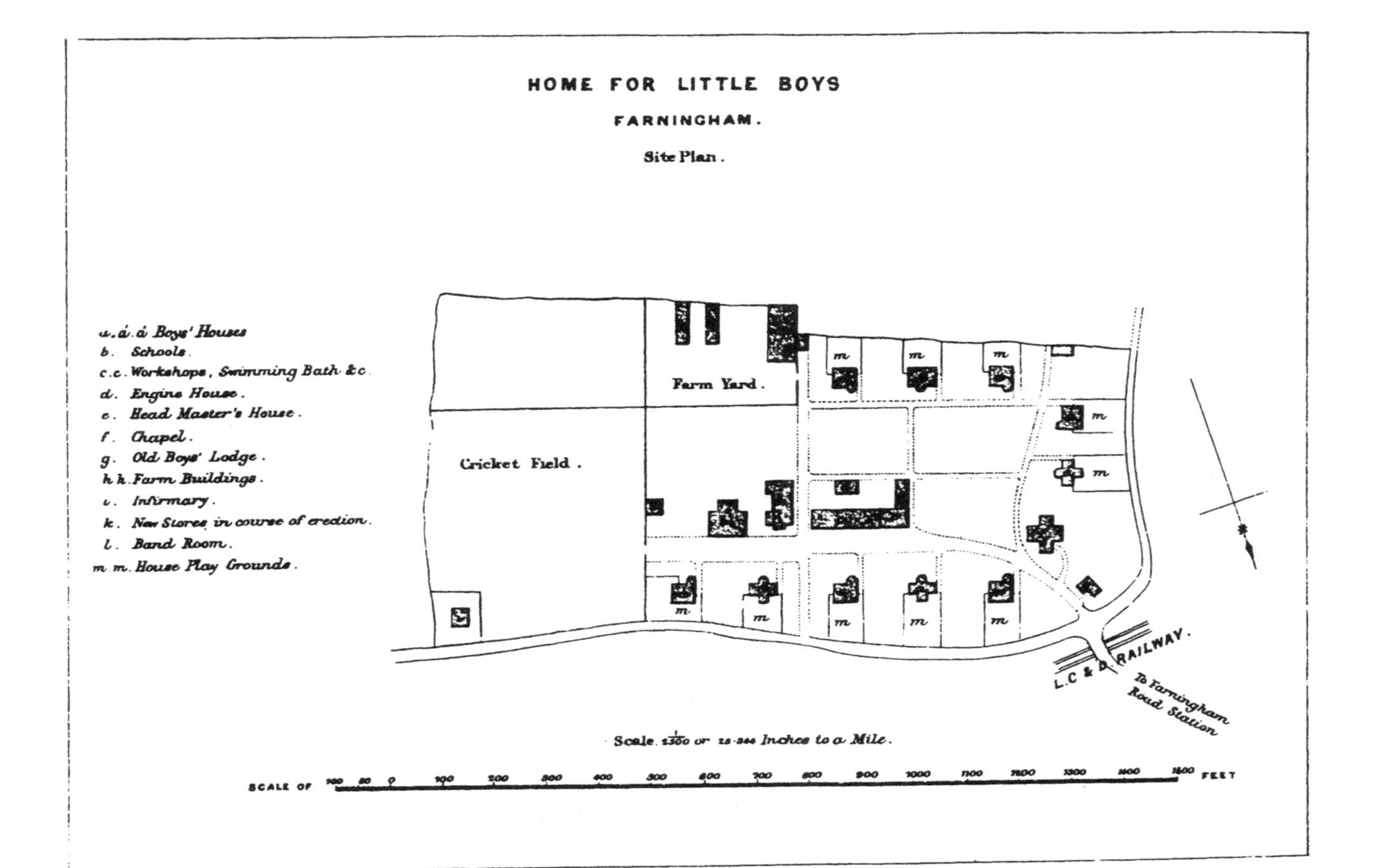
HOME FOR LITTLE BOYS
FARNINGHAM.
Site Plan.
a. á. á. Boys' Houses
b. Schools.
c. c. Workshops, Swimming Bath &c.
d. Engine House.
e. Head Master's House.
f. Chapel.
g. Old Boys' Lodge.
h. h. Farm Buildings.
i. Infirmary.
k. New Stores in course of erection.
l. Band Room.
m. m. House Play Grounds.
Cricket Field.
Farm Yard.
m
L.C & D. RAILWAY.
To Farningham Road Station
Scale 1/2500 or 25·344 Inches to a Mile.
SCALE OF
100 50 0 100 200 300 400 500 600 700 800 900 1000 1100 1200 1300 1400 1500
FEET

Fig. 6.2 (*cont.*)

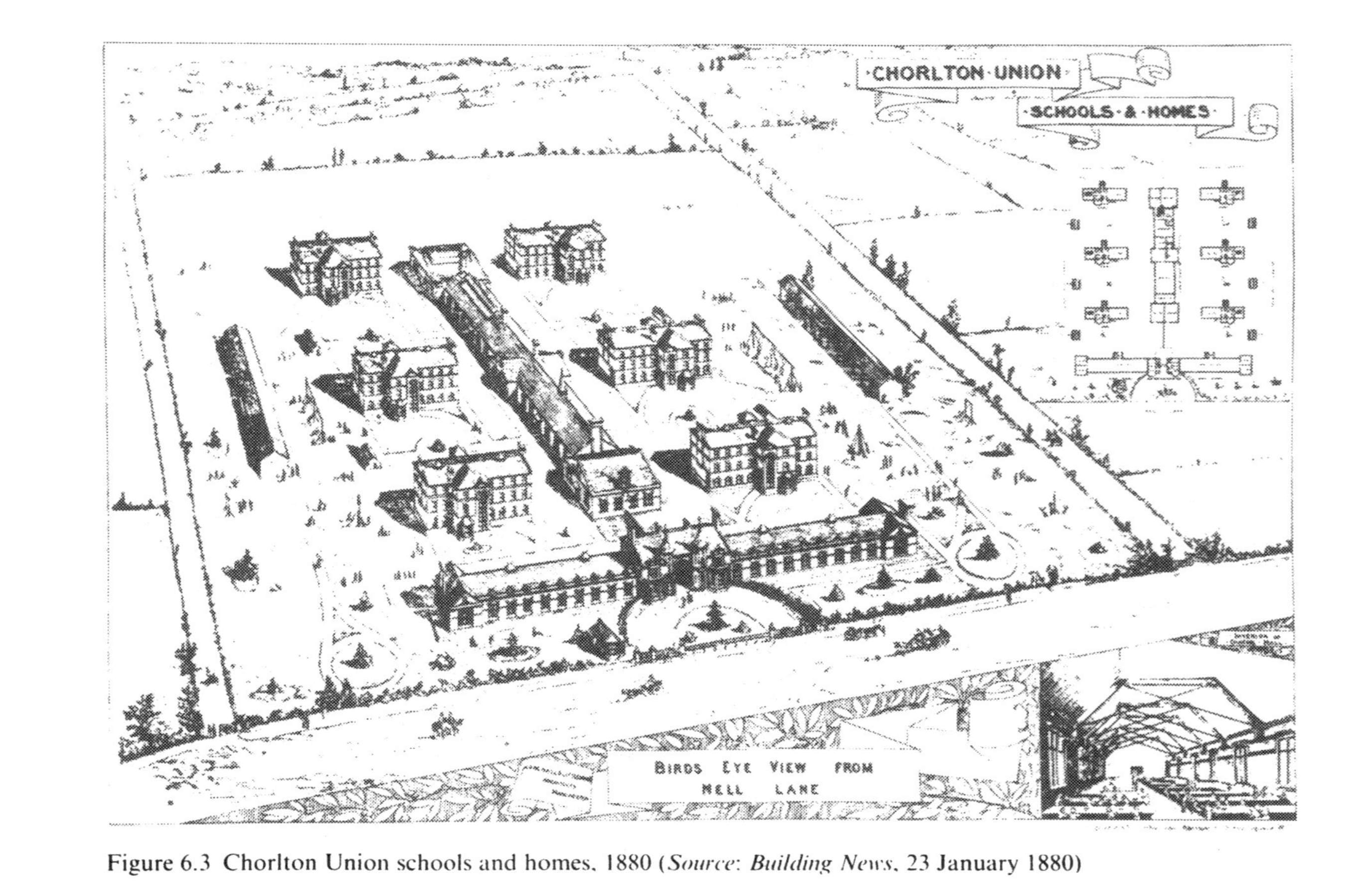


Figure 6.3 Chorlton Union schools and homes, 1880 (*Source*: *Building News*, 23 January 1880)

a single site (cf. Figure 6.3). The Banstead homes of the Kensington and Chelsea Union, for example, consisted of twenty separate houses, a chapel, a school, an infirmary, baths and a 'village shop'.[31] In one sense, such cottage homes surpassed even Mettray in their attempt to emulate the family model, as they catered for girls as well as boys, and provided substitute 'mothers' as well as 'fathers'. Yet even the Banstead 'grouped homes' were criticised for being too unnatural, too far from the family model itself. As Florence Hill observed, 'the cottage homes and their families were too large for reality; their studied neatness and affected homeliness did not give the *feeling* of home, for its genuine conditions were wanting'.[32] Critics of the 'grouped home' system recommended several alternatives, including 'scattered' homes, dispersed amongst ordinary residential districts, and boarding the children out in ordinary working-class homes, grafted onto 'normal' families. Towards the end of the century, the cottage homes policy was itself to become increasingly discredited, with boarding-out (appropriately regulated) being the preferred option. However, for the critics of the district school policy themselves, this represented an extension rather than a rejection of the 'family principle'.[33]

It is clear from this discussion that debates over Poor Law policy need to be seen in a number of broad contexts, intellectual, administrative and social. Contemporary debates over provision for pauper children were influenced by ideas of a more general provenance. The appeal of the 'family principle', for example, extended across many different fields of social policy and social science. At an institutional level, the Social Science Association provided (from 1857) an ideal vehicle for the promotion of the principle in a variety of contexts; at a conceptual level, changing attitudes towards juvenile delinquency and industrial training allowed connections to be forged between apparently disparate spheres of social concern. What also emerges from this discussion is the readiness of policy-makers and their critics to frame principles of institutional policy in very abstract terms. The family principle, for instance, was conceived as a general strategy of design, substituting the cottage for the barracks as the model of pauper training. In one important sense, however, the family principle did not depart from the logic of the new Poor Law; there still, at the heart of these new techniques of moral training, was the figure of the 'independent labourer', as prominent within contemporary discourses of social policy as he had been in 1834.

## Insane classifications: lunatics in the workhouse

It is much easier to make a distribution of the patients into distinct classes, than to construct an edifice that would correspond with that distribution, and in every respect be adapted to fulfil its intended object.[34]

I have suggested that the question of classification lay at the heart of the discourse of workhouse policy under the new Poor Law. There is no better illustration of this than the history of debate over the treatment of insane paupers. Before 1850, workhouse officials received little official guidance on the kind of regime they were to provide for inmates certified as of unsound mind', although the Poor Law Commissioners at one point suggested that chronic cases might be concentrated in district pauper asylums.[35] The official workhouse regulations initially made no specific reference to the classification of insane paupers (although local officers were required to register any paupers of unsound mind in their indoor relief lists).[36] On those occasions when the Commissioners did raise the issue with local authorities, it was often subsumed within a more general case for the building of a new workhouse. And there was no general inquiry into the treatment of insane paupers to match those that did take place during the 1840s concerning pauper children and the sick poor.

The 1834 Act permitted the maintenance of insane paupers in workhouses, as long as they were considered to be harmless; those certified as 'dangerous' could only be temporarily retained at the workhouse for a maximum of two weeks prior to removal to an asylum. The criterion for removal was thus predominantly managerial rather than medical, in the sense that it depended on the capacity of the workhouse authorities to prevent the pauper being a 'danger' either to himself or to others.[37] Under the new Poor Law, it soon became apparent that Guardians were reluctant to commit all but the most difficult cases, as the marginal cost of keeping an additional pauper in the workhouse was generally reckoned to be far less than the fees charged by asylum authorities. This cost differential had a clear impact on the direction of local policy in many districts, particularly given the increasing significance of asylum bills within overall relief expenditure. In England and Wales as a whole, the proportion of total expenditure accounted for by the maintenance of paupers in asylums rose from a mere 6% in 1857 to 14% in 1885.[38] Until at least 1874, when central government instituted a special capitation grant for the maintenance of pauper lunatics in asylums (evening out the differential between asylum and workhouse costs), the Guardians had a direct financial interest in keeping as many insane paupers as possible out of the asylum.

Figure 6.4 shows that, over the period between 1859 and 1884, about a quarter of the total number of paupers certified as insane were inmates of workhouses. The dramatic increase in the number of lunatics in asylums during the 1860s and 1870s was not, as one might expect, accompanied by a reduction in the role of the workhouse; on the contrary, absolute numbers of insane inmates in workhouses increased, while the proportion of the total remained about the same. These aggregate patterns hide distinct regional variations. In some Unions (especially in the industrial North), the workhouse remained the most common destination for the insane poor; in others (notably in Wales) non-institutional forms of maintenance were particularly

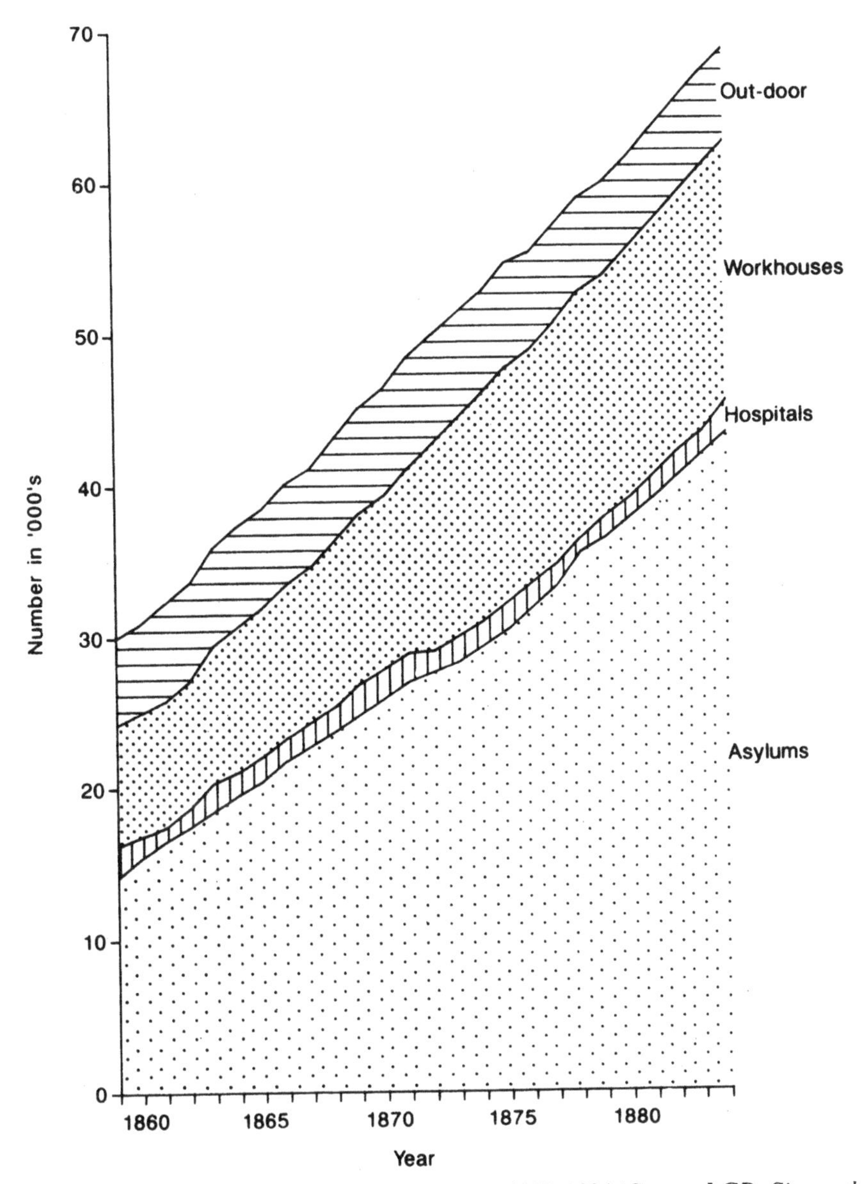


Figure 6.4 The distribution of insane paupers, 1859–1884 (*Source*: LGB, *Sixteenth Annual Report*, 1887, Appendix C62)

prevalent. In addition, the willingness of Medical Officers to certify 'insanity' may itself have varied geographically, as well as through time. In 1859, for example, the Poor Law Board reported that no less than 48 different descriptions of insanity had been used in the returns compiled by the Poor Law Union clerks; these ranged from 'Head affection after fever', through 'cerebral excitement', to simply 'silly'. The most common were dementia, idiocy and imbecility, forms of insanity usually classified as chronic or congenital.[39] Nevertheless, cases of mania and melancholia, classes more usually found in asylums, were also recorded.[40]

The retention of large numbers of insane paupers in workhouses gave rise to considerable controversy. From the first, the advocates of specialised asylum treatment had rejected other institutions, including workhouses and prisons, as inappropriate places for the care of the insane. In his campaign for the construction of pauper lunatic asylums, Samuel Tuke had described the treatment of the insane in workhouses as 'truly deplorable'.[41] In 1858, John Bucknill, editor of the prestigious *Asylum Journal*, was to repeat the same charge: in workhouses, he claimed, 'skilful management of the insane is impossible [and] the promiscuous association of pauperism and insanity is in every way objectionable'.[42] Such views as these were given added authority by the appointment of a new public authority, the Commissioners in Lunacy, whose powers of inspection were extended in 1845 to workhouses, prisons and asylums throughout the country.

The Commissioners in Lunacy saw themselves as ambassadors of enlightenment and expertise, the professional representatives of a tradition of progressive reform. Their attitude to the Poor Law authorities was critical from the start. They claimed that the treatment of the insane was no matter for those (like the Poor Law Board or the Guardians) whose judgement was too often dictated by considerations of short-term cost rather than medical treatment. It was for this reason, they argued, that so many 'harmless' insane paupers were retained in the workhouse rather than being sent to an asylum, so depriving them of a regime specifically designed to treat their needs. In the workhouse, therefore, insane paupers were abandoned to the hopeless world of Poor Law discipline:

> Placed in a gloomy and comfortless room, deprived of free exercise in the open air, and wanting substantial nutriment sufficient to promote restoration, they pass their lives in a moody, listless, unhealthy, inactive state, which is fatal to their chance of ultimate recovery.[43]

At the local level, the Commissioners in Lunacy used their powers of inspection to urge on Boards of Guardians the need to remove all curable cases from the workhouse to an asylum, as well as improvements in the care of those who remained. These improvements might include such steps as the appointment of special attendants, the provision of more nutritious dietaries, the encouragement of out-door exercise and the brightening up of workhouse wards. At the national level, the Commissioners campaigned for an extension of the asylum system so that, ultimately, all cases of insanity could be committed to specialist institutions. Their criticism of the Poor Law authorities became a matter for public debate in 1859, with the publication of a special appendix to their Annual Report, detailing their complaints about the condition of the insane in workhouses. This report, which amounted to an all-out assault on the workhouse system itself, seems to have been provoked by the increasing number of separate wards for the insane being built by Boards of Guardians. The Commissioners contended that such apparent improvements were motivated by considerations of short-term savings to the Guardians rather than long-term benefits to their insane poor. Dismissing the enhanced supervision and treatment which these new wards might allow, the Commissioners claimed that the extension of the system would effectively destroy the prospect of removing further insane paupers to asylums. Whereas the central Poor Law authorities were, on the whole, favourable to the development of new 'classified' wards staffed by specialised officers, the Lunacy authorities insisted that they would never compare with asylums. They thus rejected the evolving strategy of classification within the workhouse, arguing instead for classification without the workhouse; in other words, the entire removal of insane paupers from the jurisdiction of the Poor Law authorities. In place of the workhouse wards, the Commissioners recommended the establishment of district asylums for cases of chronic and congenital insanity, with existing asylums being reserved for the potentially curable and the truly violent.[44]

The publication of the Lunacy Commissioners' Report threatened to cause a major rift between the two central departments responsible for the care of pauper lunatics. In the event, however, the Commissioners' case was considerably weakened by the pressures that were developing within the asylum system itself. Concern at overcrowding in the asylums led to the insertion of a special clause into the 1862 Lunacy Act, allowing for the removal of chronic lunatics from asylums into workhouse lunatic wards, under certain conditions. In the same year, a Parliamentary return indicated that 115 Boards of Guardians (out of 650) claimed to provide special wards for lunatics, although by no means all of these wards were separate from the main workhouse buildings.[45] Many of the larger wards were located in urban Unions, especially those of London and the industrial North, where Guardians were becoming increasingly concerned at the escalating cost of maintaining lunatic paupers in asylums. The inexorable increase in the number of pauper lunatics was placing the asylum system under considerable stress, so that new asylums were being filled as soon as they were opened. A variety of alternative solutions were canvassed, including boarding-out and the building of small cottage-homes for the insane,[46] but none promised sufficient relief for the overcrowded asylums. The construction of two huge asylums for the chronic pauper insane by the Metropolitan Asylums Board eased some of the pressure on asylums in the London area and beyond,[47] but failed to stem the flow of insane paupers into workhouses more generally. By the late 1860s, it was abundantly clear that the crisis of accommodation within the asylums had put paid to any hope of eliminating the use of the workhouse as a receptacle for the insane.[48] In 1867, nearly half the Unions in the industrial areas of Lancashire and Yorkshire were recorded as providing substantial separate lunatic wards at their workhouses. Further such wards were built in the 1870s and 1880s, some of them representing considerable financial commitments on the part of the local Guardians. In Bradford, for example, the Guardians spent £14,000 in 1879 on new wards for their chronic insane, comprising eight dormitories, six day-rooms, one probationary ward and two padded cells.[49]

These disputes over institutional provision for insane paupers illustrate some of the ambiguities of the concept of classification. Local Guardians appear to have resorted to the building of separate lunatic wards primarily for financial reasons. When they resorted to arguments about 'classification', this was not necessarily an acknowledgement of the special needs of insane paupers; it was as much an argument for greater order within the workhouse as anything else. The central Poor Law authority also regarded the separation of insane paupers as beneficial for sane as well as insane paupers, since (in the words of one Poor Law Inspector) 'one imbecile is often able to render life in a workhouse ward distressing and painful to the rest of its occupants'.[50] The Commissioners in Lunacy, on the other hand, feared that the growth in the number of separate lunatic wards would create new barriers to the extension of the asylum system and, by implication, the improved treatment of the insane. Their doubts about the workhouse lunatic wards extended even to a revised view of the merits of separation. Whereas they had in 1847 complained at the lack of classification in most workhouses, the insane paupers being 'dispersed throughout the different buildings',[51] the Commissioners began in the late 1860s to emphasise the therapeutic advantages of a certain amount of association between sane and insane inmates. Dr Campbell, for example, described the sharing of workhouse day-rooms as 'preferable to the plan usually adopted of concentrating a number of idiots together and affording them no good example or inducement to improvement'.[52] Such comments flew in the face of advice given by Poor Law inspectors, for whom classification had become almost synonymous with separation. In the new circumstances of the 1860s and 1870s, however, a minority of alienists found still more positive things to say about the workhouse. Acknowledging its manifest inferiority as a curative environment, the President of the Medico-Psychological Association was nevertheless prepared to argue that aged, imbecile and demented lunatics prefer the workhouse to county asylums, partly from the greater freedom from discipline (from enforced order and cleanliness, baths, etc.) which they enjoy, partly from the association with sane persons there instead of the insane, and partly because it is situated nearer their own parish and family.[53]

Such a view can hardly have been popular amongst the readership of the organ of the asylum professionals, the *Asylum Journal*. It was also strangely at odds with the changing strategy of the Poor Law Board, at that moment being propelled towards more segregation within the workhouse, not less. Above all, perhaps, it reminds us of an absence at the heart of so many debates over classification – the voice of the classified.

# 7

# The politics of territory: the anti-Poor Law movement

The implementation of the new Poor Law had explosive political consequences. What had originally been represented as an enlightened measure of reform came increasingly to be portrayed outside Parliament as an assault on the rights of the poor. Goaded by influential sections of the press, and aided by a small number of radical representatives in Parliament, the popular struggle against the new law rose to national prominence during 1837. The anti-Poor Law campaigners argued that the 1834 reform heralded a dramatic transformation in the form and functions of the central state. Even apologists for the new law acknowledged its political implications, one describing it as 'essentially a system of control and restriction'.[1] To the Whig government, meanwhile, the spectre of mass resistance to the law threatened the very basis of rule from Westminster. The struggle over the new Poor Law thus raised wider questions of authority, legitimacy and, most urgently of all, public order.

Although its presence was felt throughout the country, the heartlands of the anti-Poor Law movement lay in the towns and villages of the industrial North.[2] This chapter outlines the politics of Poor Law implementation in industrial Yorkshire and Lancashire during 1837–8. The Poor Law Commissioners' attempt to introduce the new system was greeted there by a campaign of mass resistance which was most effective in localities where popular radicalism was already strong. As a result, the central authorities were forced to re-formulate their initial strategies. On the one hand, the Commissioners ceded more autonomy to the new local authorities; but, on the other, the Home Office took special measures to strengthen the position of pro-Poor Law forces in the most recalcitrant areas. This latter policy required the mobilisation of troops and police; in the celebrated case of Huddersfield, examined here, it also involved a broader recomposition of the local state.

## The popular politics of the 1830s

One is puzzled to know what to make of some of the scenes which occurred during the Oastler era. Those torch-light meetings, for instance. What mad things they look to us just now, in our cool historic moments. It is impossible to sympathise with them, and equally impossible wholly to condemn them.[3]

The intensity of popular opposition to the new Poor Law has often been remarked upon. The very violence of its rhetoric has led some historians to treat the anti-Poor Law movement as a series of spontaneous outbursts, lacking coherence or direction. Mark Hovell, for example, portrayed opposition to the new Poor Law as a reactionary campaign of 'extreme vehemence and violence [appealing] not to reason, but to passion and sentiment'.[4] There is an implicit contrast here with Chartism, which Hovell represented as a fully-fledged national movement, advancing an unambiguously progressive political programme based on a rational diagnosis of social ills. With its peculiar brand of diffuse, insurrectionary politics, and its apparent nostalgia for a mythical paternalist past, the anti-Poor Law movement seemed to Hovell to be everything that Chartism was not. However, in the light of more recent research at the grass-roots level of popular politics, such an easy distinction is no longer tenable. Beyond the ideological differences between individual leaders (say, Richard Oastler and Francis Place), there was a considerable degree of overlap between the two movements in terms of their tactics of mobilisation and their constituencies of support; indeed, recent historians have tended to interpret Chartism more as an extension than a rejection of the anti-Poor Law struggle.[5]

This chapter places considerable emphasis on the various links between radical politics and the anti-Poor Law movement. This is not to imply that affiliation to the anti-Poor Law movement automatically implied involvement in other radical campaigns, even at grass-roots level; there were clearly local variations in the extent of overlap with Chartism, for example.[6] Yet popular opposition to the new Poor Law does need to be seen in the more general context of popular agitation in Britain during the Chartist era. At an ideological level, the anti-Poor Law movement was guided by (and reinforced) a wider diagnosis of the changing role of the state, a diagnosis that was to have a profound influence on Chartism. At a strategic level, it borrowed the tactics and organisational forms of existing radical campaigns, such as the factory movement, renewing and reworking them in the process. And at a cultural level, it was sustained by a dense network of working-class institutions, from the public house to the political union.

The ideological context of anti-Poor Law protest in the industrial North was shaped by popular radical responses to the denial of working-class suffrage in 1832 and the policies of the Whig government thereafter. The same is true of Chartism, which in recent years has been portrayed less as a sudden irruption of class consciousness and more as the product of an encounter between existing traditions of radical thought and the political developments of the 1830s.[7] The exclusion of the working classes from the political nation in 1832 provided the *raison d'être* for the Chartist campaigns of six years later; critically, however, it was the character of the 'reforming' government in the intervening years which gave Chartist rhetoric much of its force and flavour. The popular radicals of the late 1830s discerned in such policies as Poor Law reform, the restriction of trades unions, Irish coercion, proposals for a new police and support for the factory system, a menacing transformation in the scale and character of state power. Thus, what had seemed a temporary setback in 1832 had become (by the time of the implementation of the new Poor Law in the industrial North) an historic turning point in the relations between the state and the people. According to their radical critics, the industrial middle classes had effectively sunk their differences with the *ancien régime* in 1832 and were bringing new, more calculating methods of oppression with them into government. There was no piece of legislation which fitted this vision more perfectly than the 1834 Poor Law.

Opposition to the central state during the 1830s was not confined to popular radicals. Indeed, the critique of centralisation has sometimes been described as one of the distinguishing features of 'Tory radicalism', a concept that has loomed large in many histories of the anti-Poor Law movement.[8] Richard Oastler, probably its most famous representative, regarded centralisation as a profound challenge to the traditions of English communal life; in 1835, he complained that 'the constitution of England knows nothing of such modern trash'.[9] Oastler's moral paternalism was certainly influential within the anti-Poor Law movement; its rhetoric of rights and duties cast the new Poor Law as an attack on the moral economy of the local community.[10] Yet popular radicals went further than Oastler in their critique of the class nature of the 1834 reform. During the campaigns of 1837, one described it as a 'weapon of Capital to oppress and grind down the honest labourer':

> The pretext is that they will raise the independent labourer. Independent indeed! There is no such thing as an independent labourer. Their object is to reduce us all to the Irish level and the machinery is calculated for it.[11]

The anti-Poor Law movement should not be seen as a single-issue campaign, waiting to be absorbed into a wider political movement. Its rhetoric implicated more general moral, economic and political assumptions, not only about the new state, but also about natural justice, local autonomy and industrial capitalism. The new Poor Law, the factory system and the proposed new police were often portrayed together as 'a series of oppressions by a middle-class government ... [designed] to subjugate the minds and bodies of the people'.[12] Each policy complemented the other; just as the factory system would depress wages, radicals argued, so the new Poor Law would keep them low:

> The existing factory and pauper systems ... appear to be working together with a harmony of effect not dissimilar to what we sometimes see in real machinery, where the teeth of a particular mill-wheel are nicely adjusted to facilitate the grinding operations of another.[13]

It was the critique of the new Poor Law, above all else, which fixed this vision of orchestrated oppression in the popular imagination. 'Chartism was the outcome of the anti-Poor Law movement', writes one historian, 'not because it was at root a social phenomenon, but because the anti-Poor Law movement was at root a political one'.[14] In strategic terms, too, the anti-Poor Law movement played an important role, by quite literally demonstrating the dramatic power of popular protest. The mass anti-Poor Law protests of 1837–8 mobilised large numbers of people throughout the industrial North, effectively marking the inaugural phase of Chartism. Many of these anti-Poor Law protests, from mass meetings on the Moors to the burning of effigies of Poor Law Commissioners, had a ritual, staged quality; their message was as much in the performance as it was in the script.[15]

The symbolism of much anti-Poor Law protest did not diminish its violent, confrontational character. After all, it was the popular campaign against the Poor Law which brought the Metropolitan police and the military onto the streets of towns like Bradford, Huddersfield and Oldham during 1837–8. In a sense, such a forceful response pointed up the weakness of government rather than its strength. Episodes of popular violence, including the blockading of Guardians' Board Rooms, assaults on Assistant Poor Law Commissioners and concerted attacks on workhouses demonstrated the effectiveness of popular pressure at a *local* level. There, the symbols of the new law (individuals and institutions) were readily identifiable – and vulnerable too, as the forces of local authority were quite unable (in all but the largest towns) to protect themselves without the help of the military. The billeting of troops was a last resort, and a most inflexible one at that. The military commanders of the Northern district were convinced of the need to secure permanent barracks, minimising the distance troops would have to travel to likely centres of disorder (Figure 7.1). More generally, as on previous occasions, what the authorities feared most was a rash of simultaneous outbreaks in a number of different districts. As Mather has pointed out,

> The threat to English society in the Chartist period did not in fact arise from the strength of the resistance which the rioters were capable of offering to the forces of the crown, but from the tendency of disturbances to occur almost simultaneously in different places.[16]

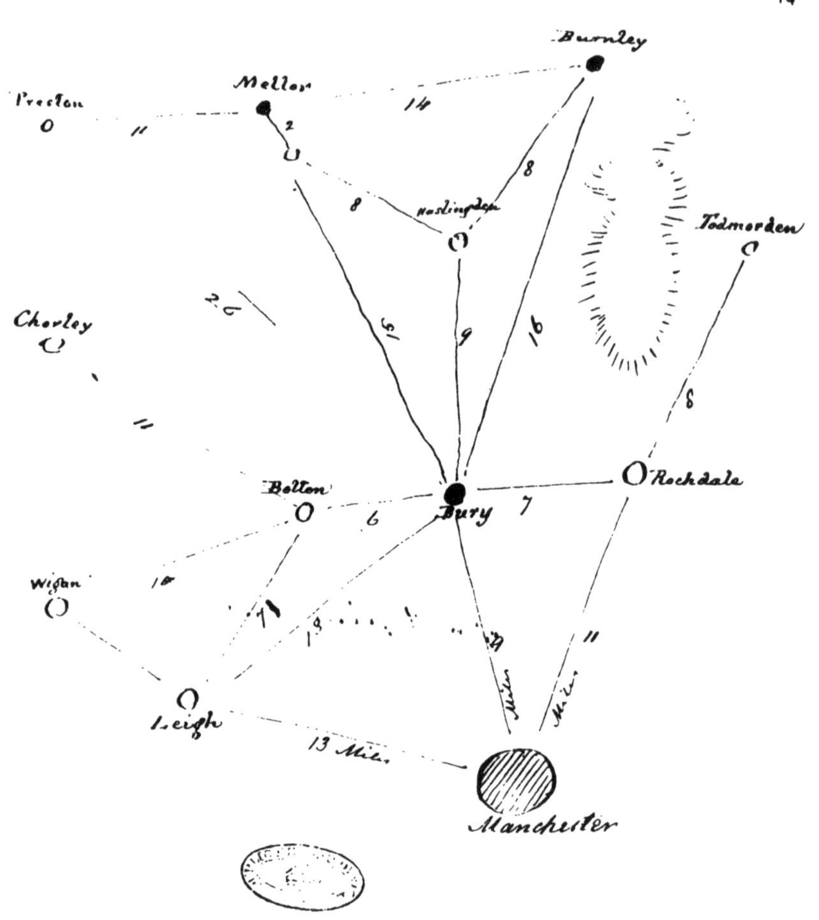


Figure 7.1 Proposed locations for military barracks, 1841 (*Source*: C. Napier to S. Phillipps, 2 March 1841, HO 45/41)

Although dispersed, such disturbances were not necessarily unco-ordinated. During the late 1830s, for example, the radical press played an important role in anti-Poor Law politics, giving advance publicity and detailed accounts of demonstrations in villages and towns throughout the country. The *Northern Star*'s calls for the mobilisation of popular force were particularly unambiguous; as it warned in February 1838 (after the defeat of a Parliamentary motion for repeal of the 1834 law), 'We must repeal the law *out* of the House'.[17]

It was at the local level that anti-Poor Law protest achieved its most notable successes. This was not simply a by-product of the inefficiency of the central state, for the very lifeblood of the movement was drawn from the

Figure 7.2 The New Poor Law Bill: England's infernal machine (*Source*: S. Roberts, *The peers, the people and the poor*, 1838)

sphere of local politics. In places like Huddersfield and Oldham, the organisation and strategy of anti-Poor Law protest, emerged almost organically from those of other radical movements, such as the factory campaign. Anti-Poor Law broadsheets and handbills exploited a rich vein of popular political culture and commentary. In *Give it a fair trial*, a satirical tract distributed in the Huddersfield area during 1837, for example, an alternative 'Poor Law' reform was proposed for the 'amendment' of the wealthy classes. The ladies and gentlemen of the aristocracy were to be incarcerated in separate apartments of 'Union palaces', meeting each other 'through a golden grate', just 'once a week'.[18]

The movement against the new Poor Law was a genuinely popular campaign, involving vast numbers of ordinary people, as paupers, handloom weavers, factory-hands, agricultural labourers, overseers and ratepayers. Anti-Poor Law protest brought politics into the everyday lives of these people, creating a momentum which threatened to overwhelm the forces of order. In a sense, this violent clash between 'high' and 'low' politics was a direct consequence of the extension of central authority envisaged by the reformers of 1834. The revolution in Poor Law government was often represented by anti-Poor Law campaigners as an encroachment into previously autonomous areas of local life. The territorial imagery in Figure 7.2 was not, therefore, merely metaphorical. As Eileen Yeo has argued, working-class politics during the late 1830s was to a large extent a politics of space, a contest for the control of the household, the street, the factory, the community, the vestry, the meeting-house and the pub.[19] The scale of the mobilisation of armed force in the towns and villages of the industrial North during the late 1830s, and the large number of more mundane local conflicts over rights of access and assembly, suggest that popular politics had indeed become a struggle for territory.

**Region in crisis: Poor Law politics in the industrial North**

The implementation of the new law in the industrial North was delayed until the Autumn of 1836, when the process of Union formation in the rural South was almost complete (see pp. 37–42). In tactical terms, there was much to be said for the Commissioners' cautious, region-by-region approach; yet circumstances conspired to undermine their strategy. The woollen trade in the West Riding was about to enter a prolonged recession, intensified by the collapse of American demand for English cloth, and the brief recovery of 1837–8 was followed by a still more severe downturn affecting the Northern industrial economy as a whole. The gathering economic crisis was compounded by a growing sense of political frustration. As Richard Oastler observed in May 1837,

> Whilst . . . labourers in one of the biggest hives of industry in the world are starving in the streets, whilst our capitalists and the richest merchants in the universe are trembling lest the arrival of an American packet should disappoint all their hopes and anticipations, . . . whilst embarrassment, like a hurricane of despair, is driving the nation to destruction – what are they doing in Parliament? Talking about bubbles and quarrelling about straws![20]

This sense of crisis in the industrial North deepened the growing divisions between popular radicals and Whig reformers. Differences over factory legislation, trades unions, suffrage reform and Ireland had already made their mark, and the new Poor Law antagonised popular radicals still further. In October 1836, Halifax radicals upstaged a Whig reform dinner by inviting Feargus O'Connor to a large meeting in the local theatre, where the new Poor Law was roundly condemned. A month later, at a meeting in the Union Street Methodist chapel in Oldham, the factory campaigner J. R. Stephens spoke of a new 'system of anti-social competition'; 'the connection between this cursed Poor Law bill and the factory system', he argued, was 'intricate and inseparable'.[21]

Figure 7.3 Alfred Power, 1880 (*Source*: *New Monthly Magazine*)

Given this coalescence of specific grievances within a generalised sense of social crisis in the industrial North, the appointment of Alfred Power as the Assistant Poor Law Commissioner to the region was perhaps unfortunate (Figure 7.3). Power was no stranger to the industrial North; in 1833, as a representative of a Royal Commission on factory labour, he had attracted the wrath of short-time campaigners.[22] In 1837, the haste with which he set about introducing the new system again drew criticism. In a matter of weeks, he established a large number of Unions in Lancashire and the West Riding, ready for the Guardians' elections in March. Power attempted to disarm his critics by emphasising the necessity of consultation. As he assured the Commissioners,

> My answer to all such attacks is – Nothing is yet done ... No regulations are yet issued to *this* Union. I am here to confer with your representatives on the materials before us, and to take care that every information is laid before the Commissioners of any peculiar circumstances that may exist recommending a departure from the usual course, before a single order is issued.[23]

But the anti-Poor Law campaigners scorned Power's diplomacy. In Richard Oastler's words,

> Power pulls a nice face and says, 'The Guardians have nothing to fear – the Commissioners will issue nothing which the Guardians would disapprove of'. There is no clause in the Act for this. He is a liar![24]

The Guardians' elections in March 1837 provided the first comprehensive test of the strength of opposition in the industrial North. Anti-Poor Law committees had already sprung up at local and Union levels (many of them based on existing short-time committees), and these were later supplemented by county-level organisations. Their campaigns achieved several notable successes; at Todmorden and Oldham, for example, no elections took place. In those Unions where elections were held, the outcome was mixed; one historian observes that 'the larger towns and cities returned Liberal Guardians, while rural areas were usually solidly Tory'.[25] This pattern does not necessarily reflect the geography of anti-Poor Law feeling, not least because Guardians' elections were conducted on a comparatively restricted franchise; the property qualification and multiple voting system significantly reduced the direct impact of popular opinion. Furthermore, party affiliation, even where it can be established, was not in every case a reliable indicator of Guardians' attitudes towards the new law. Much depended on local political circumstances. Strong support for anti-Poor Law Guardians was particularly evident in such places as Huddersfield, Bolton, Bradford and Keighley.

The results of the Guardians' elections caused considerable alarm within the Poor Law Commission and amongst supporters of the Whig government. Even Alfred Power himself acknowledged that 'a considerable number of respectable and influential persons' now opposed the new system.[26] In May 1837, a large anti-Poor Law demonstration was held at Hartshead Moor, between Huddersfield and Bradford. Huge crowds[27] converged on the Moor from all over the West Riding, carrying a large number of colourful banners. The demonstration was not merely about the rights and wrongs of Poor Law reform; it signalled a wider process of popular political mobilisation. As if to underline this, one banner displayed two slogans, the first reading 'NO POOR LAW AMENDMENT BILL – NO BASTILE PUNISHMENTS', the second 'UNIVERSAL SUFFRAGE – VOTE BY BALLOT'. Some speakers (including Oastler) concentrated their demands on the repeal of the Poor Law; but several (notably the future Chartists, O'Connor and O'Brien) demanded broader schemes of parliamentary reform. The meeting was widely reported in the regional and national press, and even the sceptical Power was forced to acknowledge that it was 'more than a Huddersfield affair' [28]

The Hartshead Moor demonstration gave the anti-Poor Law campaign new momentum at the regional level; yet it was at the local level that the struggle remained most effective. In Huddersfield, for example, the new law was successfully obstructed throughout 1837. Events in Huddersfield had much more than local significance; indeed, by June national attention was focussed on the struggle there. Richard Oastler himself stood as an anti-Poor Law candidate at two Parliamentary elections in the town within as many months. Although he was narrowly defeated on both occasions, Oastler's campaigns raised political temperatures still further. The second election, held at the end of July, was a particularly violent affair; on polling day in Huddersfield, a contingent of Metropolitan police, armed with cutlasses, fought a pitched battle with a crowd numbered at between twenty and thirty thousand.[29] In the end, military force was called in to restore order. These disturbances were followed by more violence at Wakefield a few days later, and both events contributed towards a growing sense of political crisis in the industrial districts as a whole. To one historian, indeed, it 'seemed as if the West Riding was on the verge of insurrection'.[30]

The political implications of anti-Poor Law protest caused concern amongst local supporters of the Whig government, who feared that the Tories would exploit the issue at the impending general election. In June, a delegation of West Riding Whig Members of Parliament attended the Poor Law Commission, urging some delay in implementation, at least until economic circumstances had improved.[31] It was the political factor which seems to have weighed most heavily with leading Ministers; the Home Secretary, John Russell, was particularly concerned at the effect of the Poor Law issue on the precarious position of the government in the Commons.[32] Two weeks after the Parliamentary delegation had been received at Somerset

House, Russell intervened. Calling for a delay in implementation, he advised the Commissioners that 'extended enquiry, longer experience and frequent discussion' would extinguish the flames of anti-Poor Law protest.[33] This well-publicised gesture effectively stalled the progress of the new system in the most hostile areas. Although the delay was only temporary, it strengthened the hand of anti-Poor Law Guardians in places such as Huddersfield, Bradford, Bury and Ashton.[34] During the autumn of 1837, the focus of agitation moved temporarily to Bradford. In November, a violent confrontation took place between a military force and anti-Poor Law campaigners demonstrating outside a Guardians' meeting. While Alfred Power blamed local agitators, the radical M.P. John Fielden claimed in Parliament that the new Poor Law was being enforced 'at the point of the bayonet'.[35] Fielden himself was active in the formation of the new South Lancashire Anti-Poor Law Association. And from the end of November, the newly-founded *Northern Star* was devoting considerable space to the anti-Poor Law struggle.

The events of the winter of 1837–8 mark a key moment in the anti-Poor Law campaign. In many Unions, the Commissioners had succeeded in establishing the basic machinery of the new Poor Law; clerks had been elected, administrative sub-districts established and new officials appointed. (This was the minimum requirement for the 1836 Registration Act, based on the machinery of the new Poor Law.) Russell's concessions and the crushing defeat of Fielden's anti-Poor Law motion in the Commons (in February 1838) together signal something of a turning point. Yet anti-Poor Law protest was not entirely extinguished; indeed, the Guardians' elections of March 1838 resulted in a strengthening of resistance in the most recalcitrant areas.[36] In Oldham and Todmorden, the elections were effectively boycotted, and in Huddersfield, the anti-Poor Law movement actually increased its representation on the Board; as one pro-Poor Law Guardian there commented, 'we have more opposition within and less without'.[37] Subsequently, anti-Poor Law campaigners in Huddersfield used the 1834 Act itself to challenge the legitimacy of Guardians' elections in several townships.[38] By the autumn, when the outcome of the disputed elections had become clear, anti-Poor Law Guardians had gained effective control.[39] Only in April 1839 did responsibility for the administration of relief pass from local townships to the Huddersfield Board of Guardians; and, even then, anti-Poor Law feeling continued to influence the direction of local policy (see chapters 8 and 9).

Russell's famous 'concessions' of 1837 formed but one side of a broader counter-offensive in the Poor Law struggle. Alongside these public gestures of conciliation, he co-ordinated a private campaign to neutralise opposition in the heartlands of anti-Poor Law protest. This campaign was centred on Huddersfield, the focal point of anti-Poor Law politics throughout 1837.

**Local politics: anti-Poor Law protest in Huddersfield**

The strength of local resistance to the new Poor Law in the industrial North varied considerably. In some areas, the new system was put into operation with almost no obstruction whatsoever; in others, implementation was stalled for years. Any interpretation of this uneven geography of resistance must consider a variety of factors. The degree of popular political mobilisation prior to the introduction of the law was important, as was the disposition of local elites. Local political ecology could also be significant, especially in the context of relations between local township authorities – such as the vestries – and the new Boards of Guardians In this section, I examine the local politics of resistance in the Huddersfield district, one of the heartlands of the anti-Poor Law movement.

It is clear that the relative effectiveness of anti-Poor Law protest in Huddersfield, as in Oldham, reflected a sophisticated tradition of local radicalism.[40] The influence of the Huddersfield radicals was felt far beyond the town itself; men such as Lawrence Pitkethly and William Stocks, for example, played an active part in radical politics at regional and national levels. In 1838, Pitkethly described Huddersfield as the 'centre of an extensive and extending sphere of [political] operations'.[41] The town and its political hinterland, which was dotted with industrial villages, provided the anti-Poor Law movement with a huge reservoir of popular support. Henry Vincent, a political missionary from the London Working Men's Association, was particularly impressed by the strength of local radicalism. In a letter written from Huddersfield in 1838, he reported to his friends that

> You have no idea of the intensity of radical opinions here. You have an index from the numerous public house signs – full length portraits of Hunt – holding in his hand scrowls [sic] containing the words Universal Suffrage, Annual Parliaments and The Ballot. Paine and Cobbett also figure occasionally.[42]

This remarkable statement locates local radicalism not only within a broader political tradition, but also within the institutions and symbols of popular culture.

As well as offering an ideological framework for popular grievances, local radicalism provided the anti-Poor Law movement with a ready-made campaigning machine. The Huddersfield radicals formed a well-organised and relatively cohesive group; cohesive enough, indeed, to be described by one Poor Law Guardian as a 'knot of sturdy democrats'.[43] The 'Pitkethly faction', as the *Leeds Mercury* disparagingly referred to them,[44] had extensive experience in organising political societies, defending trades unions, publishing radical newspapers and running specific campaigns. The short-time factory committees provided one power-base for the anti-Poor Law movement; local government provided another, particularly where open

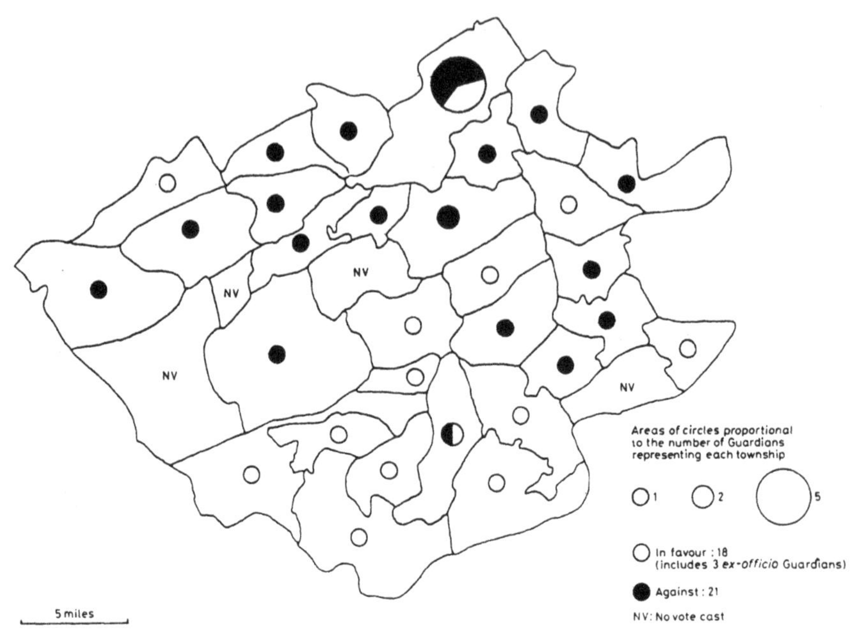


Figure 7.4 The Huddersfield Guardians' vote on the election of a clerk, 3 April 1837 (*Source*: HGM. See Figure 8.2 for base map).

vestries survived. Although historians have often portrayed Huddersfield as almost the pocket borough of the local Whig magnate,[45] the political hegemony of the Ramsden family was never absolute. By the mid 1830s, radicals had gained considerable influence in the Huddersfield vestry, being elected as overseers, surveyors and constables.[46] Their grip on the vestry strengthened during the anti-Poor Law campaign, the workhouse board elected in 1838 being dominated by local radicals. Deprived of a role in other local institutions, radicals saw the open vestry as a stronghold of popular democracy. As elsewhere in the industrial North, and most notably in Oldham, township meetings and open vestries provided 'points of access to the local political process'.[47]

By the Autumn of 1836, when the first stirrings of anti-Poor Law protest were being felt in the industrial North, radicals had thus established a significant political base in Huddersfield and its surrounding district. In January 1837, a few days after Alfred Power had paid his first visit to the town, the radicals held a large torch-lit demonstration, at which Richard Oastler and Feargus O'Connor were the main speakers. One of the many banners visible on this occasion read *Those whom God hath joined together, let no devil of a Commissioner put asunder. If the paupers and the poor rates are to be Commissioned, so should the land and the landlords!*[48] Agitation intensified as the Guardians' elections drew near, with local radicals speaking at demonstrations throughout the new Poor Law Union.[49] The agitation had its desired effect; the Guardians elected in March 1837 refused even to appoint a clerk. Alfred Power, who was present at their first meeting, regarded the decision as a direct blow against the new law; as he reported to the Commissioners, 'many of the Guardians had come pledged from the townships they represent to support an adjournment of the proceedings in a hostile spirit'.[50] Figure 7.4 shows the way the Guardians voted on the election of a clerk, indicating a clear division between the Southern part of the Union (around Holmfirth), which was represented by predominantly pro-Poor Law Guardians, and the Northern industrial townships around Huddersfield itself, which voted solidly against any appointment. The township of Huddersfield itself, which returned five Guardians, was divided. This pattern suggests that anti-Poor Law feeling was most effectively mobilised in Huddersfield and its neighbouring townships, contradicting the common assumption that resistance was concentrated in the rural hinterlands of industrial Unions.[51]

The effectiveness of anti-Poor Law resistance in Huddersfield depended to some extent on the support of many local Tories. Tory–radical co-operation had already become a feature of popular politics in Huddersfield during the 1830s, under the auspices of the short-time campaign. Such alliances developed only fleetingly and haltingly in places like Huddersfield; indeed, I have argued elsewhere that radical co-operation with local Tories was tactically motivated and only partially successful.[52] Nevertheless, the Huddersfield radicals certainly supported the Tory Richard Oastler during his two election campaigns of 1837. This support was premised on Oastler's pledge to 'tear the accursed new Poor Law from the statute book', as well as on his local pedigree; in contrast, Edward Ellice, his Whig opponent at the May contest, was an outsider.[53] The Poor Law struggle dominated both elections, with leading anti-Poor Law activists (including Stephens, Bull and Pitkethly) joining Oastler on the hustings. The pattern of voting at the May election, as revealed by analysis of poll books (Table 7.1), provides an insight into the social basis of the Tory–radical alliance. Support for Oastler was clearly at its strongest in two social groups: traditional Tory voters (such as farmers and magistrates), on the one hand, and the more humble occupational groups (including publicans and tailors) on the other. In contrast, Ellice received substantial support amongst the manufacturing and professional classes. Nevertheless, his majority was much reduced, suggesting that radical tactics (which included exclusive dealing)[54] had nearly paid off.

Oastler's role in the Huddersfield anti-Poor Law agitation was important; but it should not be exaggerated. Indeed, on more than one occasion during the struggles of 1837, his ability to control his 'followers' was shown to be distinctly fragile. His disavowal of the radical principles of universal suffrage and secret ballots during his election campaigns, for example, cannot have

Table 7.1. *Voting in the Huddersfield election, 6 May 1837*

| | OASTLER | (%) | ELLICE | (%) |
|---|---|---|---|---|
| Gentlemen | 7 | (100) | 0 | (0) |
| Farmers | 51 | (88) | 7 | (12) |
| Woolstaplers | 21 | (60) | 14 | (40) |
| Publicans | 42 | (58) | 30 | (42) |
| Tailors, drapers | 30 | (52) | 28 | (48) |
| Other retailers | 37 | (45) | 45 | (55) |
| Service trades | 17 | (39) | 27 | (61) |
| Merchants, manufacturers | 55 | (32) | 117 | (68) |
| Professionals | 19 | (29) | 46 | (71) |
| Other | 11 | (30) | 26 | (70) |
| Total | 290 | (46) | 340 | (54) |

*Source:* Huddersfield poll book, Institute of Historical Research

endeared him to the hustings crowds.[55] Nor did his pleas for restraint at anti-Poor Law demonstrations meet with much success. On 5 June 1837, for example, a Guardians' meeting in the Huddersfield poorhouse was disrupted by a large crowd of men and women. The meeting was only resumed when more secure premises were found in the centre of Huddersfield, at the Albion Tavern. Yet more violence was to follow; the Chairman had not even finished reading over the minutes of the previous meeting, when a large stone was hurled through one of the hotel windows. A huge crowd had gathered outside the building, threatening to overwhelm even the anti-Poor Law campaigners (including Oastler) who were in attendance. One of the Tory magistrates, an *ex-officio* member of the Board, addressing the crowd from the window, pledged that he would not read the Riot Act (despite being requested to do so by pro-Poor Law Guardians) so long as there was no further violence. To popular delight, the Guardians again refused to appoint a clerk, and in the night of celebration that followed, effigies of the Poor Law Commissioners were put to flame in the marketplace.[56]

These dramatic developments underline not only the powerlessness of the local pro-Poor Law faction, but also the key role of the Tory magistracy. The events of June 1837 convinced local Whigs that, without further intervention from central government, the new Poor Law could not be implemented in Huddersfield. From this point on, their local struggle became an issue of truly national concern. Just as anti-Poor Law campaigners saw the Huddersfield Union as perhaps their most important stronghold, the government came to regard the local struggle as a critical test of their authority. This was the place where the other, less publicised side of Russell's strategy was to bear fruit.

**Reconstituting the local state: the magistracy**

You must recollect that the power of a magistrate does not begin and end with sending a sheep-stealer to gaol. (Russell to Melbourne, 9 October 1835)[57]

In the eyes of the central authorities, the magistracy had a number of important roles to play in the implementation of the new Poor Law. As influential figures within the local community, they would be amongst the first to be consulted by Assistant Commissioners prior to the formation of Unions. As *ex-officio* members, their influence on local Boards of Guardians could be of critical importance, particularly in the early years of the new system. And as Justices of the Peace, they had a special duty to maintain order and suppress violent disturbances. Events in Huddersfield during 1837 were to show that these roles were not necessarily complementary. Having failed to secure the wholehearted assistance of the local Bench during 1837, the central authorities eventually determined on an alternative strategy: it was time to appoint new magistrates more sympathetic to the new law.

In the summer of 1837, there were five acting Magistrates in the Huddersfield division of the West Riding Commission of the Peace, an area coterminous with the new Poor Law Union (Table 7.2). The five played an influential role in the local Tory community, their interests bound together not only by political loyalties, but also by marriage contracts, property deals and industrial concerns.[58] Their Tory credentials did not imply abstention from the commercial world; on the contrary, most were involved in significant industrial undertakings. Joseph Armitage, for example, owned a large complex of woollen mills at Milnsbridge, while Sir John Lister Kaye was active in developing mines and canals on his estates near Huddersfield. This conjunction of Tory politics and industrial interests was not as peculiar as it may seem to the historian looking for neat correspondences between parties and interests; indeed, it has recently been suggested that Tories formed a majority of those magistrates with industrial interests in the West Riding during this period.[59]

Tory domination of the Huddersfield Bench was clearly a source of irritation to local Whigs, and irritation turned to outright hostility during 1837. Following the repeated failure of the new Board of Guardians to proceed with the election of a clerk, Whig Guardians from Holmfirth complained to the central authorities that the local magistracy were failing to implement the new law. Immediately after the debacle of 5 June (described above), the local postmaster reported that 'the mob has triumphed, the law is set at nought', while another Whig, John Sutcliffe (a local factory-owner) begged Joseph Hume (a supporter of the 1834 Act and an associate of

Table 7.2. *The Huddersfield magistracy in 1837*

| Name | Residence | Date of oath |
|---|---|---|
| *Acting magistrates in June 1837* | | |
| Joseph Walker | Lascelles, Hall, Lepton | January 1829 |
| John L. Kaye | Denby Grange | July 1830 |
| Joseph Armitage | Milnsbridge House, Huddersfield | April 1833 |
| B. N. R. Batty | Almondbury | April 1833 |
| W. W. Battye | Mold Green, Dalton | March 1837 |
| *Additions to the Commission in November 1837* | | |
| William Brook | Mold Green, Dalton | December 1837 |
| Joseph Starkey | Kirkheaton | December 1837 |
| John Sutcliffe | Huddersfield | March 1838 |
| Thomas Starkey | Huddersfield | April 1838 |
| George Armitage | Kirklees Hall | September 1840 |
| Joseph Brook | Huddersfield | May 1846 |
| John Starkey | Huddersfield | January 1852 |
| John Brooke | Armitage Bridge | January 1852 |

*Sources:* Names of Acting Magistrates, QD1/108 (WYRO); Fiat Book, West Riding of Yorkshire, C234/44 (PRO); Dedimus Books, C193/46–7 (PRO).

Edward Ellice, influential father of the Huddersfield M.P.) to exert his influence with the government. Yet another letter, almost identical to Sutcliffe's, found its way directly to the Home Office: 'We are only short of two things here', it complained, 'and that was some soldiers and the creation of about four Magistrates of liberal opinions'.[60]

While they were privately campaigning for the appointment of new magistrates, the Huddersfield Whigs (like their colleagues elsewhere in the North) were publicly urging the central authorities to take a more diplomatic approach to Poor Law reform in the industrial districts. It was in an attempt to defuse this developing political crisis that John Russell had recommended a temporary delay in the further implementation of the new law. However, this well-publicised gesture was accompanied by more secretive manoeuvres. Russell's first concern as Home Secretary was with the apparent breakdown in local authority. Appalled by reports of events at Huddersfield, he immediately ordered both troops and metropolitan police to the town. In addition, he censured the local magistrates for both their failure to read the Riot Act and their reluctance to call in outside support.[61] This criticism was a prelude to a new strategy, the outcome of which would only become apparent many months later. Russell made his intentions clear in a letter written in July to Earl Harewood, the (Tory) Lord Lieutenant of the West Riding:

> The Magistrates acting for Huddersfield have been so far under the influence of the prevailing terror that the property, the freedom of action, and even the lives of the peaceable inhabitants of Huddersfield are without secure protection for the future, unless some addition is made to the Magistracy in that part of the West Riding.[62]

The opportunity to make new appointments to the Huddersfield Bench arose only a few weeks later. As was customary following a General Election, the Lord Lieutenant had prepared a list of recommended additions to the West Riding Commission (including several names for the Huddersfield division) for the Lord Chancellor. It appears, however, that Russell was not satisfied with these suggestions. In consultation with Alfred Power (the Assistant Poor Law Commissioner) and W. Stansfield (the new Whig M.P. for Huddersfield), he submitted an additional list of four names to the Lord Chancellor, candidly describing their appointment as 'essential for the due enforcement of the Poor Law Amendment Act'.[63] In the present context, it is less the fact of Russell's intervention – which was not unprecedented – than the motives that inspired it which are significant. Power's influence in the matter, the subject of much speculation amongst anti-Poor Law campaigners, does appear to have been important. In naming his choice of new magistrates, Russell informed Harewood that 'Mr Power says they are deserving our trust, and I hope there is no objection'.[64] Stansfield, meanwhile, was painfully aware of the political costs of the anti-Poor Law agitation, clearly visible in the reduced Whig majority at Huddersfield and elsewhere.[65] This directly electoral factor was, as we have seen, also an important consideration for Russell.

The eventual outcome of the behind-the-scenes negotiations of autumn 1837 was the addition of no less than eight names to the Huddersfield division of the West Riding Commission of the Peace in November 1837, four of them Power's nominees and four Harewood's (Table 7.2).[66] However, addition to the Commission was not sufficient for individuals to act as magistrates; to proceed to the Bench, nominees had to take a qualifying oath – the *Dedimus Postatem* – at the Quarter Sessions.[67] It appears that Power's nominees were the first to take this oath, two in December 1837, and two the following spring. All four were prominent local factory-owners and advocates of the new Poor Law, and all had appeared on the Whig (and pro-Poor Law) side of the hustings during the 1837 borough elections.[68] There could be little doubt, therefore, of Russell's intention in securing their appointment. It was not simply to broaden the social composition of the magistracy (one of his longer-term aims), since industrial and commercial interests were already well represented on the Bench. It was rather to secure a change in the balance of local political forces which had hitherto obstructed the implementation of the new Poor Law.

The first fruits of Russell's intervention were realised in Huddersfield at the beginning of 1838. At a meeting held in the Huddersfield Court House, chaired by one of the new magistrates, the Guardians finally resolved to appoint a clerk, so setting the new Poor Law in motion. While a majority of Guardians voted against any appointment taking place, the Commissioners contended that three votes cast in favour of any candidate was sufficient for that candidate to be appointed. One of the newly-appointed magistrates immediately informed the Home Secretary of the result, describing the proceedings as 'a triumph of Law and Order'.[69] The anti-Poor Law campaigners interpreted the result in diametrically opposite terms, as a negation of the principle of local democracy; in Oastler's words, 'it is not we that oppose the law; it is the Commissioners and their tools'.[70] Denouncing what he described as a 'system of French centralisation', Oastler directed his fire at Alfred Power, his old adversary: 'I am resolved to know if your masters in Somerset House are hereafter to govern the BENCH as well as the paupers and the Guardians'.[71]

## Conclusion

Anti-Poor Law protest in the industrial North was sustained by a campaign of mass resistance which appears to have taken the central authorities almost by surprise. In those localities where popular radicalism was already influential, the movement succeeded in stalling the implementation of the new law. The marked success of anti-Poor Law protest in Huddersfield during 1837 reflected the considerable influence and organisation of the local radicals, as well as the tactical alliance with Oastler and the Tories. Both radical and Tory opponents of the new Poor Law exploited the rhetoric of local autonomy; the spectre of centralisation ('one of those queer words used nowadays')[72] provided them with a common focus. Russell's direct intervention in the appointment of new magistrates, as we have seen, ultimately ensured that the new Poor Law was at least introduced at Huddersfield. Nevertheless, the echoes of these struggles were to reverberate within the new system for years to come.

# 8

# From township to Union? The geography of Poor Law administration at a local level

> In no Union in England is the township feeling so strong as it is in Huddersfield. Every township in the Union, except those included in the Borough, is a separate local government district; every township has its own rate-collector; and every township looks upon its neighbour as its natural enemy.
>
> J. S. Davy, Local Government Board Inspector (1882)[1]

Although resistance to the new Poor Law was particularly intense in the Huddersfield area, it ultimately failed to prevent the introduction of a new system of local administration. Thirty-four separate townships were incorporated within a single Union authority, responsible for the direction of relief policy and the appointment of local officials. This chapter considers the long-term significance of this apparent transformation, from a predominantly local perspective. Clearly, the history of the new Poor Law in Huddersfield is not simply a miniature version of national, or even regional histories. The relative intensity of anti-Poor Law protest, and the unusually large extent of the Union, for example, made a significant difference locally. Yet it would be quite wrong to divorce the local experience of the new Poor Law from its broader context. What we are concerned with in this chapter is less an isolated locality than a national system seen from local perspective. Once Poor Law administration was lifted out of the domain of the parish or the township, the definition of 'local' authority was to remain perpetually in dispute. The study that follows, therefore, is less concerned with the local relief system *per se* than with the relationships between administrators in the locality and policy-makers from beyond.

## Pauperism and the local environment

The Huddersfield Poor Law Union formed in January 1837 was one of the largest in the country, serving thirty-four townships and a population of around 100,000. Its physical geography was dominated by the Colne and Holme rivers, flowing towards Huddersfield township in the north from the

Pennine moors in the west and the south, respectively. These steep-sided river valleys were dotted with numerous small settlements; indeed, in 1831 less than a quarter of the population of the future Union lay in Huddersfield township itself. Its large and expanding population depended increasingly on local industry. In 1795, Huddersfield itself was described as 'peculiarly the creation of the woollen manufactory, whereby it has been raised from an inconsiderable place to a great degree of prosperity and population'.[2] The town's weekly market, at which fancy goods were a particular speciality, attracted hundreds of manufacturers from the surrounding district. Huddersfield operated as the regional distribution centre for both raw wool and finished products; at the hub of sub-regional road and canal networks, it linked the district with the whole of the industrial North.

The fortunes of the entire Huddersfield area were thus closely tied to the local woollen industry. Times of prosperity saw the building of houses, canals and mills; times of economic hardship, on the other hand, resulted in large-scale distress. Such fluctuations were endemic to the regional economies of the industrial North; as a Huddersfield surgeon noted in 1818, 'it needs but a slight acquaintance with history to discover that the prosperity of towns and districts, like states, is fleeting and transitory'.[3] The rhythms of the industrial economy made their mark in the Huddersfield district during the 1820s. Following a period of fierce competition and wage cuts, the woollen industry was plunged into a sudden crisis in 1826, when a financial panic resulted in bankruptcies and unemployment throughout the district. This crisis had a clear impact on poor relief provision, particularly in the fancy-weaving areas to the southeast of Huddersfield proper. In Huddersfield township itself, relief expenditure rose by 150% in a single year; while expenditure in the townships of the future Poor Law Union almost doubled (Figure 8.1). Unemployment persisted throughout 1826; in Huddersfield, a special relief fund was set up for the weavers, and in neighbouring townships labourers were set to work on the roads. This experience had a far-reaching effect; local radicals, such as William Stocks, complained that the prospect of complete destitution had forced weavers to accept lower wages.[4] The crisis of 1826 measured the extent to which the local economy had become tied to a much wider economic system.

The economic effects of unexpected hardship were to some extent mitigated by the existence of a large number of friendly societies and sick clubs in the Huddersfield area. These self-help institutions provided limited financial support for subscribers (and their families) in times of ill health, unemployment or bereavement. Their role in sustaining the poor through periods of temporary crisis should not be forgotten.[5] However, these sources were inevitably inadequate in times of mass unemployment, as in 1826. Furthermore, there is little evidence to suggest that friendly societies played any part in supporting the chronically sick, the elderly, orphans or the disabled;

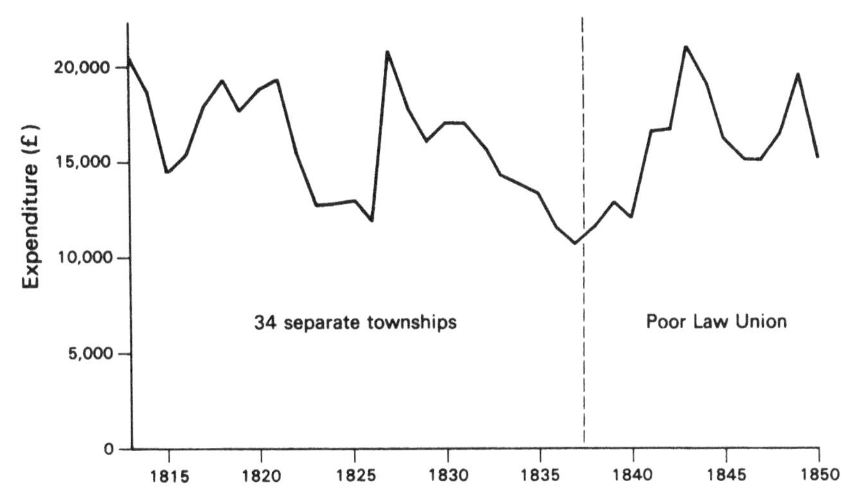


Figure 8.1 Poor Relief expenditure in the Huddersfield district, 1813–1850 (*Source*: Annual Poor Rate Returns)

indeed, these cases were usually deliberately excluded from membership.[6] Outside times of heavy unemployment, such groups dominated local overseers' relief lists. In Thurstonland during the late 1830s, for example, the 'permanent' parish poor consisted of sixteen paupers described as either 'decrepit' or 'infirm', two blind people, two lunatics, two families with disabled children and two single mothers.[7] Such a list was broadly typical, with due allowance for the size of the community, of most townships in the industrial North during times of healthy employment.[8] There is little evidence of relief of able-bodied men on anything but a short-term basis, and no sign of the widespread adoption of the allowance or roundsmen systems which were condemned by the Poor Law reformers. The Huddersfield overseers, for example, claimed in 1834 that they never gave relief simply 'to help low wages'.[9]

The evidence, uneven and incomplete as it is,[10] suggests that relief provision in the Huddersfield area before the arrival of the new Poor Law mainly focussed on the non-able bodied. Only in moments of acute crisis (such as unemployment, the birth of a child, sickness or a death in the family) would the able-bodied poor seek poor relief, as a last resort. Such relief would not necessarily take the form of a dole; it might include the payment of doctors' bills or the provision of materials (including food, blankets, clothes, shoes or coffins). In some townships, able-bodied men were required to work on the roads. There is little evidence of local adoption of the workhouse test, designed to deter able-bodied applicants. Although about one-third of the townships in the intended Union provided some form of poorhouse in the mid 1830s, their use was limited and their size generally small (see pp. 147–8). Apart from the Huddersfield workhouse, most of them would not have been distinguishable from neighbouring buildings, in either appearance or internal layout. The Honley workhouse constructed in 1783, for example, was a plain two-storey house with only a few beds.[11] Even in Huddersfield itself, there was a general reluctance to use the workhouse as a deterrent. Workhouse populations were dominated by the elderly and by children. In 1833, 48 of the 106 inmates of the Huddersfield workhouse were children; in nearby Golcar, half the 18 inmates were children.[12]

Under the old system, therefore, local overseers were overwhelmingly preoccupied with the relief of elderly paupers, the care of orphan children, the treatment of lunatics and the maintenance of deserted wives. A large proportion of their business was taken up with protecting the financial interests of their township. Much attention, for example, was paid to the possibility of removing paupers to their parish of settlement, or placing orphans out as 'apprentices'.[13] Such tasks placed a considerable burden on local administrators, providing an important stimulus for the adoption of Sturges Bourne's Vestry Act of 1818, which was particularly popular in the industrial North. By 1835, about two-thirds of the townships in the Huddersfield area had appointed salaried overseers, and about a third had elected select vestries under this Act.[14] Paid or unpaid, the local overseer was frequently the most important administrator in each township, responsible not only for the relief of the poor, but also for such diverse tasks as the enumeration of population, the compilation of militia lists and the capture of vermin.

## Centralism and localism, 1838–1884

The coming of the new Poor Law promised major changes in the geography of local relief administration. Although the Commissioners acknowledged the relative efficiency of local authorities in the North,[15] they insisted on substituting the Union for the township as the fundamental unit of Poor Law administration. They hoped to replace hordes of part-time, locally appointed officers with a professional corps of staff under the control of the Union as a whole. In this strategy of geographical concentration, the role of the Board of Guardians was to be critical. For, in theory, it was to provide a new voice in local affairs – the voice of the Union.

### *The Board of Guardians: the voice of the Union?*

The amalgamation of small townships within a larger authority was intended to lift administration beyond the realms of 'local influence' (see pp. 37–41).

The property qualification for elected Guardians, the multiple voting system and the appointment of *ex-officio* members were together intended to protect the general interests of property against those of any particular group or locality. In practice, however, the new local authorities were far from disinterested bodies. Guardians were frequently to find themselves at loggerheads over Poor Law policy, partly as a result of broader political differences. Their office itself presented considerable opportunities for the exercise of influence,[16] reason enough for their deliberations to be closely reported in the local press.

The thirty-four townships of the Huddersfield Union (Figure 8.2) were initially represented by forty-one elected Guardians; by the end of our period (1884), the number had been increased to an unwieldy seventy, to allow for population changes. Occupationally, they were a diverse group, with manufacturers and farmers dominating, and gentry, shopkeepers and professionals also being represented.[17] The significance of these distinctions can be exaggerated; divisions on the Board were as likely to reflect the importance of local loyalties, social standing or political affiliation. Most Guardians regarded themselves as township delegates rather than servants of the Union. As on many other Boards in the region, 'the voice of the township was both loud and persistent'.[18] As long as the township remained the basic unit of chargeability for most paupers, there remained a powerful financial rationale for this localism. There was thus widespread suspicion of any proposal which involved heavy expenditure from the common fund of the Union.

During the early years of the new system, the Guardians elected for the township of Huddersfield itself were amongst the most hostile to the new Union. Resistance was co-ordinated by the Huddersfield vestry and its radical spokesmen, including William Stocks and Lawrence Pitkethly (see pp. 123–4).[19] Throughout 1839, the vestry and local Guardians refused to acknowledge the jurisdiction of the Union over the Huddersfield workhouse.[20] Although the institution was eventually handed over to the Guardians (who immediately dismissed its Chartist master), the dispute revived anti-Poor Law agitation; in the elections of March 1840, Pitkethly was elected as a Guardian in Huddersfield and Stocks was only narrowly defeated in Honley. The radical case against Union influence in local affairs foreshadowed subsequent developments. The principle of localism was institutionalised on the Board of Guardians itself, with the establishment of four sectional relief committees in February 1840. Applications for out-door relief were henceforth to be considered by the Guardians elected for the appropriate district. This devolution of authority away from the Union level was condemned by the architects of the 1834 reform; Edwin Chadwick, for example, described the principle as a 'bounty to minorities'.[21] According to one local critic, the committees created 'a mongrel sort of affair, neither the

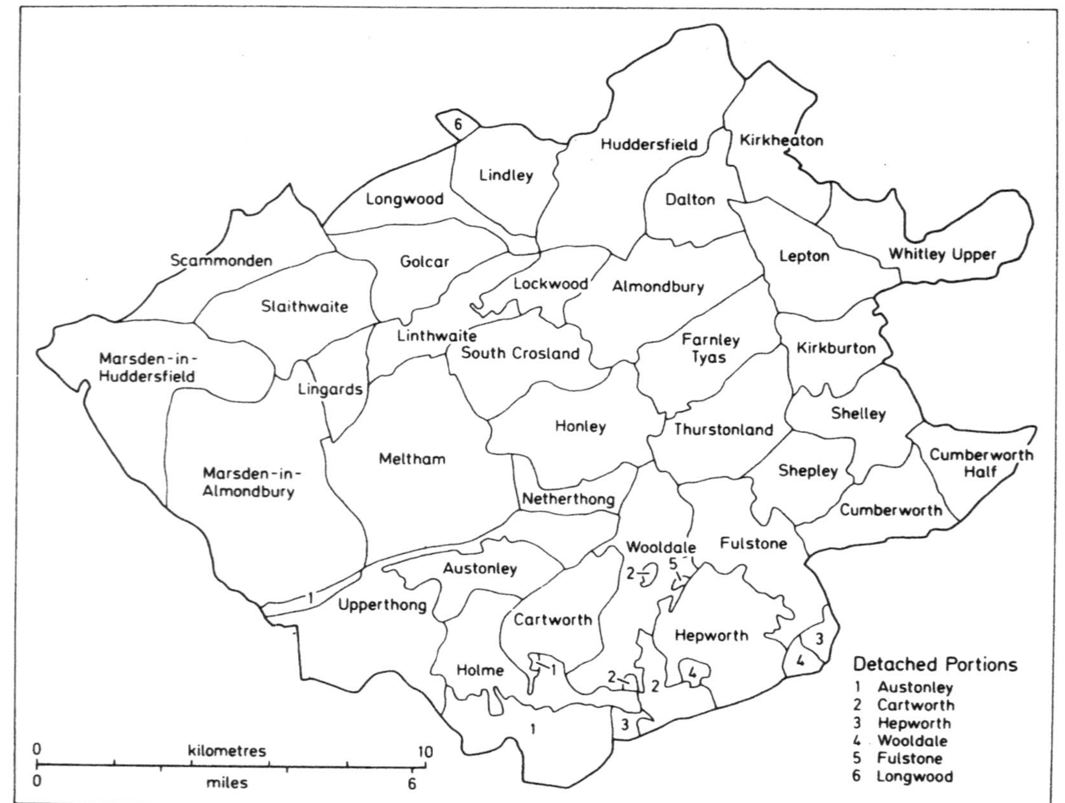


Figure 8.2 Townships of the Huddersfield Union (*Source*: MH 12/15108)

old [system] nor the new'.[22] In this respect, however, Huddersfield was far from unique; between 1847 and 1858, the central authority sanctioned the division of no less than fifty-two Boards of Guardians into sectional relief committees.[23] In Huddersfield, as an official inspector lamented in 1882, each committee was left 'practically independent in the administration of relief'.[24]

The devolution of relief administration towards the local level was also reflected in the Guardians' decision (in 1841) to hold their summer meetings fortnightly rather than weekly. The central authorities argued that more frequent meetings would provide a more 'salutary control' over relief business. Yet, as elsewhere in the industrial North, fortnightly meetings were actually extended into the winter.[25] This practice gave greater autonomy to individual Guardians, overseers and relief officers, who were empowered to grant relief in cases of urgent need. It also encouraged further delegation of the day-to-day management of local workhouses to individual Guardians and Union officers. After 1840, there were five separate workhouse committees, each containing Guardians elected for the townships in which the workhouses were situated[26] (see chapter 9).

The central authority acceeded to some degree of devolution in the administration of out-door and indoor relief in the hope that it might forestall more far-reaching challenges, especially from anti-Poor Law Guardians.[27] Throughout its history, the Huddersfield Union was perpetually under threat of dissolution, proposals for a division surfacing whenever debate amongst the Guardians had reached a critical point.[28] Arguments in favour of division can, broadly speaking, be divided into three kinds. The first, advanced by the Guardians of the Holmfirth district to the south of Huddersfield, was that the Union was simply too large and unwieldy for efficient management. This was a common complaint amongst Guardians in the larger Northern Unions.[29] The second argument, advanced by the representatives of Huddersfield and its neighbouring townships, focussed on the growing lack of proportion between township populations and their representation on the Board. The incorporation of Huddersfield and its neighbouring townships as a Borough in 1868 strengthened this case, as we shall see. The third argument, used at different times by different groups, emphasised irreconcilable differences over workhouse policy. Whenever proposals for new construction failed to gain the required consent of two-thirds of the Guardians, calls for a division of the Union inevitably followed. During the 1840s and 1850s, it was the Holmfirth Guardians who argued this case; after the mid 1860s, it was the turn of the Borough Guardians.[30]

The balance of power on the Board of Guardians altered decisively following the incorporation of Huddersfield in 1868. Frustrated by their lack of influence, the Borough Guardians co-operated with the Corporation

Table 8.1 *Huddersfield Union: borough and outlying townships, 1881–1885*

| District | Population 1881 | Population change 1841–1881 | Rateable value 1881 | Number of guardians 1881 | 1885 |
|---|---|---|---|---|---|
| Borough | 81,823 | +82% | £284,348 | 15 | 35 |
| Golcar | 29,544 | +54% | £98,358 | 9 | 13 |
| Holmfirth | 25,871 | +3% | £82,584 | 12 | 12 |
| Kirkburton | 19,560 | −1% | £54,985 | 10 | 10 |
| Total | 156,798 | +44% | £520,275 | 46 | 70 |

*Sources:* Borough Memorial, MH 12/15104; HGM, 17 April 1885.
*Note:* The *Borough District* includes Almondbury, Dalton, Huddersfield, Lindley and Lockwood; the *Golcar District* includes Golcar, Lingards, Linthwaite, Longwood, Marsden-in-Almondbury, Marsden-in-Huddersfield, Scammonden and Slaithwaite; the *Holmfirth District* includes Austonley, Cartworth, Fulstone, Hepworth, Holme, Honley, Meltham, Netherthong, Upperthong and Wooldale; the *Kirkburton District* includes Cumberworth, Cumberworth-half (Skelmanthorpe), Farnley Tyas, Kirkburton, Kirkheaton, Lepton, Shelley, Shepley, Thurstonland and Whitley Upper. See Figure 8.2 for township boundaries.

in a campaign for increased representation. They pointed to the two-to-one majority held by the outlying townships on the Board, the 'country villages ruling the roost' despite the relative shift of population and rateable value towards the Borough.[31] It appears that the central authority was moved to act only by the prospect of a divided Union, following a threat of legal action from the Corporation in March 1882.[32] The end result of central intervention – an unwieldy Board of seventy Guardians, equally balanced between the Borough and the outlying townships (Table 8.1) – suggests that the principle of representation according to rateable value (rather than simply population) had been conceded, despite official fears that ward-by-ward elections of large numbers of urban Guardians would invite 'log-rolling'.[33] The size of the new Board itself created difficulty; in May 1884, for example, proceedings in the Boardroom were halted by a dispute over seating arrangements. As one Local Government Inspector, J. S. Davy, wryly observed, 'good administration is impossible wherever the number of administrators is too great for the work to be done'.[34]

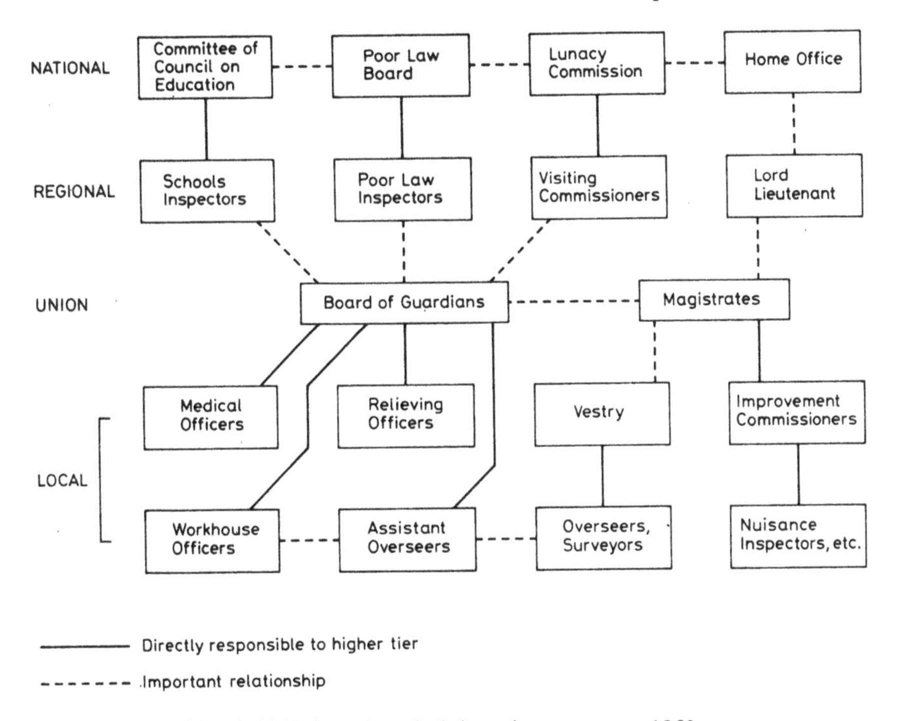


Figure 8.3 Huddersfield Union: the administrative context c. 1850

*Union officials: the geography of authority*

The 1834 Poor Law created a local bureaucracy organised around the Union rather than the township. At the centre of the new machinery was the Guardians' clerk. In addition to Poor Law duties, the clerk supervised the activities of the local registrars of births and deaths under the Registration Act. All the other salaried officials appointed by the Huddersfield Union – the relief officers, workhouse masters and matrons – were responsible only for a particular district or institution within the Union (Figure 8.3). This raises an issue about the geography of authority; to what extent did these local officials serve the interests of the Union as a whole?

The establishment of Unions was supposed to enable the replacement of a large number of part-time (township) overseers by a small number of full-time (district) officials. 'In diminishing the number of channels through which intelligence has to pass', the central authority asserted, 'we diminish in direct proportion the temptations to act partially or from any motives of a personal nature' (see chapter 3).[35] In Huddersfield, however, no less than fourteen lowly-paid relief officers were appointed in December 1838, by a predominantly anti-Poor Law Board. After the Guardians' elections of the following year, the number was reduced to four, at salaries of £100 per annum. Each of these officers was responsible for districts containing (in 1851) up to 44,000 people. It was their task not only to receive applications and distribute relief, but also to visit the poor (including non-resident paupers) and to assist in the transfer of infirm or insane paupers to and from workhouses. In times of crisis, the Guardians were forced to appoint temporary assistants, and there is some evidence that the relief officers informally employed helpers (including, in two recorded cases, their wives).[36] The relief officer responsible for the populous Huddersfield district (which until 1883 comprised four townships) was under particular pressure. Over the period as a whole, the Huddersfield district accounted for an increasing proportion of the Union's fortnightly relief bill.[37]

The pressure on the relief officers ensured that other officials continued to play a role in the granting of relief under the new Poor Law. In theory, it was only in exceptional cases of 'urgent and sudden necessity' that Guardians and overseers could distribute relief. Yet there is evidence to suggest that local township authorities in many Northern Unions played a more substantial part in the distribution of relief.[38] In September 1842, a Huddersfield overseer was reprimanded by the Guardians for his continuing involvement in the business of relief.[39] Two years later, an Assistant Commissioner was surprised to find 'self-constituted committees in almost every township who attend the relieving officer in paying the paupers'; significantly, he added that 'they have it more their own way when the Guardians meet less often'.[40] During particular emergencies, more formal arrangements were introduced. A distress committee was formed in Kirkheaton during a period of widespread distress in 1842, to supervise the employment of the able-bodied poor on local roads. Relief, in the form of wages or provisions, was funded from both the poor rate and the highway rate. In this instance, moreover, the overseer, the Guardian and the local shopkeeper were one and the same person.[41]

In addition to the relief officers, the Huddersfield Guardians appointed a large number of 'assistant overseers'. In some respects, these appointments – common throughout the industrial North – were further concessions to existing practice in an area where the rate collection and settlement business was unusually burdensome.[42] In contrast to the old system, these new officers were supposed to be responsible to the Union rather than individual townships. However, such a division of authority could prove difficult to sustain, particularly where large numbers of officers were appointed. In Huddersfield, for example, the Guardians eventually appointed an assistant overseer for each one of the thirty-four townships of the Union. This policy was (reluctantly) sanctioned in London, in recognition of the strength of local feeling. The chairman of the Guardians had warned in 1844 of 'a deeply rooted antipathy amongst the ratepayers against strangers being appointed to intermeddle with their own parochial affairs, which extends to every class of society'.[43] Assistant Commissioner Alfred Power was more blunt. In his view, too many ratepayers simply couldn't 'bear to see a stranger come into their township and take away their money'.[44]

**Patterns of local policy**

Although the Huddersfield Guardians took formal responsibility for the administration of relief in April 1839, they were given no specific instructions on policy beyond the usual guidance on the appointment of officers and the routine functioning of the Board. The position was the same in many other Unions in the industrial North, although the Commissioners felt confident enough to issue out-door labour test orders in some areas (Figure 3.7). Few of the Huddersfield Guardians, whatever their stance on relief policy, regarded the Union workhouse as a panacea for the problems of pauperism.[45] Indeed, they loudly objected to the blanket use of the workhouse test as a 'self-acting' mechanism for distinguishing between the deserving and undeserving poor. During the 1840s and 1850s, it was impossible for officials to propose the adoption of the workhouse test without raising the spectre of the 'bastile' and reviving memories of the turbulent struggles of the late 1830s.

Only in times of general economic crisis would large numbers of able-bodied men turn to the Poor Law authorities. The winters of the early 1840s, for example, brought widespread unemployment throughout the industrial districts. In the fancy-weaving area of Huddersfield, especially around Kirkheaton, relief expenditure soared, as it had during 1826 (Figure 8.1). In the West Riding as a whole, adult able-bodied pauperism reached 37% of all paupers relieved during the Quarter ending March 1843; if dependants are included, this meant that well over half those relieved were classed as able-bodied.[46] These circumstances perhaps explain the general reluctance of the Huddersfield Board (and many others in the North) to pursue a blanket policy of deterrence; indeed, when asked (at a Parliamentary inquiry in 1843) whether the Guardians ever applied the workhouse test, the Chairman replied 'No, not at all; never'.[47] Instead, they preferred to see unemployed labourers working on roads and in stone quarries.[48] In one form or another, this form of relief practice appealed to a wide range of local opinion. Some portrayed it as a local version of the Commissioners' 'labour test', while others emphasised its financial benefits to the Union. The central authority, however, was unimpressed, largely because the organisation and funding of the system was left to the township authorities themselves. As one Assistant Commissioner complained, 'their heart's desire [was] not letting their paupers work for anyone but themselves'.[49]

The continuing localism of relief practice also drew criticism from the new district auditor appointed in 1845. He was frequently to complain about the Guardians' lack of control over local overseers; in 1846, for example, he disallowed several items in overseers' accounts, including money relief granted without the consent of the Guardians, expenditure on an unauthorised poorhouse, and the payment of the Huddersfield vagrant office rent out of township funds. These disallowances (and surcharges) met with a furious response from the Guardians. Their articulate Chairman, the Reverend J. M. Maxfield, complained that the auditor's strict interpretation of the regulations 'in their literal and grammatical sense' had failed to allow for the 'peculiar situation and circumstances of the Union'; the ruling against temporary relief in money, in particular, was 'arbitrary, cruel and impracticable'. Maxfield even raised the possibility of renewed mass resistance to the new Poor Law. As he commented in a letter to *The Times*,

> We did not enter upon the administration of the law because we approved of it, but simply because it was the law of the land ... The Guardians of the Huddersfield Union have never been the tools of the Commissioners ... They will never become the slaves of an auditor.[50]

In some cases, these accounting disputes resulted in the issuing of retrospective orders sanctioning unauthorised payments; in others, there followed legal proceedings against local overseers. Either way, it was clear that the weapons of surcharge and disallowance placed the district auditor in a position of considerable power over this most recalcitrant of local authorities.

However, it was not until 1852 that the central authority felt able to extend the formal regulation of relief policy further. In August 1852, the Poor Law Board issued an Out-door Relief Regulation Order to Unions in Lancashire, the West Riding and London. Under this Order, relief to able-bodied males already in employment was forbidden, and those who were unemployed were only to receive relief in return for a task of work. In addition, a third of relief given to the out-door poor (a half in the case of the able-bodied) was to consist of relief-in-kind, and all was to be distributed at no greater than weekly intervals. Although these terms were less stringent than those imposed elsewhere (see chapter 3), Northern Guardians immediately launched a concerted campaign to have them revised. The Huddersfield Guardians condemned the Order as an 'unconstitutional infringement of the rights of local Boards'; as one of the most outspoken put it, 'the Guardians were the best judge of what was necessary, not the three despots of Somerset House'.[51] While such rhetoric was in a literal sense anachronistic (the 'three despots of Somerset House' had long been superseded by the Poor Law Board, based at Gwydir House), it had powerful resonances, reviving memories of the Chartist years. The central authority quickly retreated and issued a revised Order, giving the Guardians full discretion over the relief of the non-able bodied, and allowing them more scope in granting short-term relief to the able-bodied. It has been argued that the wording of this revised Order, and the instructional letter which accompanied it, opened the door to widespread evasion.[52] The letter placed a particularly narrow interpretation on the clauses excluding relief to working men; thus 'a man working for wages on one day and being without the next, or working half the week and being unemployed during the remainder, and being then in need of relief, is not prohibited by this article'.[53] Officials later acknowledged that this statement threatened one of the fundamental principles of 1834: the abolition of relief-in-aid-of-wages.[54]

There does not seem to have been any dramatic shift in relief policy in the Huddersfield Union following the receipt of the new Order in December 1852. Relief-in-aid-of-wages continued to be unregulated in the case of widows with children, and the elderly or infirm; even able-bodied males were occasionally exempted from the Order, where the Guardians saw fit. There is also little evidence to suggest that the Guardians applied the required labour test with anything like the stringency demanded by the central authorities, before the 1870s. Where work was required, it generally took place on lands adjoining local workhouses. According to one inspector, this constituted 'the least efficient test of destitution that can be resorted to, as the work is pleasant, and from my observation, it is rather liked by paupers than otherwise'.[55] On only one occasion during this period did the Guardians attempt a more systematic scheme. In February 1854 it was reported that hundreds of unemployed labourers were employed breaking dross.[56] When trade recovered, even this scheme was dropped.

Although there was considerable debate over local workhouse policy (discussed in chapter 9), the idea of a workhouse test remained unpopular amongst most if not all of the Huddersfield Guardians. Opponents of new workhouse provision continued to exploit the rhetoric of the 'bastile'. James Tolson, their most vociferous spokesman, took every opportunity to remind his fellow Guardians of 'the time when the Poor Law system was forced on the acceptance of this district by sword and bayonet'.[57] Edward Clayton, a radical shopkeeper, also couched his opposition to a new workhouse in the language of the 1830s. 'To his mind', it was reported in 1858, 'the cottage was as dear to the poorest of the poor, as the palace was to the noble of the land. Any law which has the tendency to destroy this love of home in the cottage was a bad law. The tendency and aim of the workhouse system was to destroy that feeling' (compare Figure 4.3).[58] Tolson and Clayton were influential figures on the Board, the former becoming Chairman in 1858, and the latter succeeding him two years later. Even the supporters of workhouse construction felt the need to distinguish their proposals from those of 1834. Insisting that 'the bastile cry was an old and exploded idea',[59] they emphasised the improved treatment, efficient management and better classification that a new workhouse would bring. As Joshua Hobson, the radical turned Tory, put it, 'There were interwoven with the early management of [the 1834] law principles which it was the duty of every Englishman to oppose. The case however is now different. These principles have been given up – abrogated' [60]

The campaign for additional workhouse provision culminated in the establishment of new workhouses at Deanhouse (Honley) in 1862 and at Crosland Moor (Huddersfield) in 1872 (see pp. 152–5). Although the new workhouses were not primarily intended to deter able-bodied paupers, their construction did allow the Guardians greater flexibility in their policies towards out-door paupers. The timing of the construction of the larger Crosland Moor institution coincided with a national campaign against out-relief. While the Huddersfield Guardians opposed more extreme proposals for the entire abolition of out-door relief, they did urge their officials to tighten their surveillance of all the cases on their books, especially where single mothers were concerned. From 1875, the Guardians adopted the increasingly common practice of publishing regular lists of all individuals receiving poor relief. At the same time, out-door relief to vagrants (in the form of tickets issued for lodging houses) was halted, and more stringent tasks of labour imposed in the new vagrant wards at Crosland Moor. In 1877–8, labour tests for other male able-bodied paupers were formally established at both workhouses. The clerk to the Union, John Hall, described these tests as 'both profitable and repressive'.[61]

The large expenditure incurred in the construction of new workhouses gave the Guardians an incentive to look for savings elsewhere. John Hall attributed the reduction in relief expenditure which took place during the 1870s to the 'workhouse test' itself.[62] On the surface, at least, his case was a powerful one. While pauperism in England and Wales as a whole fell between 1870 and 1875 by a quarter, in Huddersfield it was halved. The biggest reductions occurred in 1872–3, precisely when the Crosland Moor workhouse came into operation. As Figure 8.4 shows, the reduction was concentrated amongst the out-door poor. Yet these trends were not irreversible. After 1875, relief expenditure again began to rise, and the winters of 1878–80 saw many able-bodied adults returning to the relief rolls. The crisis once again illustrated the essentially cyclical nature of pauperism in the industrial districts. In times of high employment, male able-bodied pauperism was relatively insignificant; during periods of depression, in 1879 as in 1826, 1842 and 1862, it could increase dramatically. A similar pattern was exhibited by female pauperism, although its base level was consistently higher, even after the campaign of the 1870s. In the half-year ending December 1878, for example, there were twice as many women on out-door relief as men; and the vast majority of those relieved because of 'insufficient earnings' were women.[63]

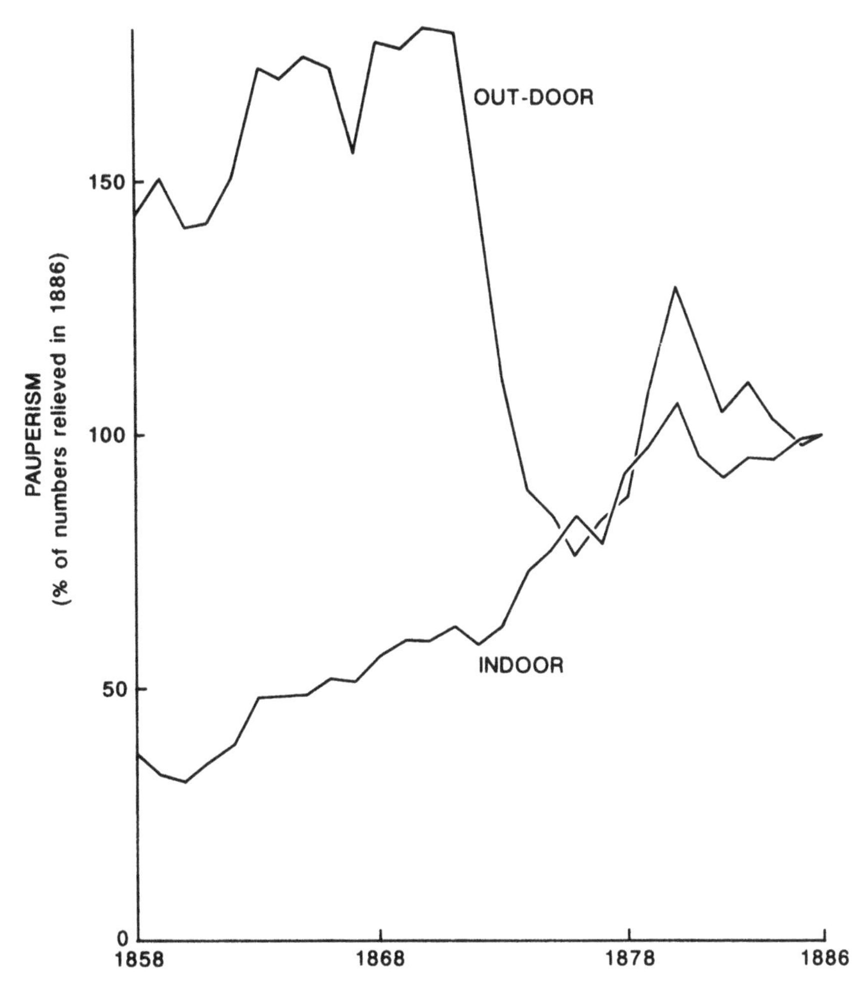


Figure 8.4 Pauperism in the Huddersfield Union, 1858–1886 (*Source*: Annual Returns of Pauperism)

**Conclusion**

The transition to the new Poor Law in the Huddersfield area created, as elsewhere, an entirely new tier of local government. In some important respects, it reduced the power of existing local authorities. The radical vestry of Huddersfield, for example, was effectively deprived of much (but by no means all) of its influence. Under the watchful eye of the district auditors (appointed after 1844), local overseers and assistant overseers were forced to fall into line with officially-sanctioned practice. Several of the Union officers

– especially the clerk and his relief officers – occupied new and distinct positions within the reformed system, without precedent under the old. Yet the impact of such changes was compromised in Huddersfield, as elsewhere in the industrial North, by the devolution of authority away from the Union level. The creation of sectional relief committees undermined the coherence of the Board as a whole, one indication of the fact that the Guardians saw themselves as the representatives of townships rather than servants of the Union. As we have seen, devolution was sanctioned by the Commissioners only in order to save the Union from dissolution. The same rationale lay behind the official response to calls for a readjustment of representation on the Board in 1882–3. In effect, therefore, 'township feeling' was incorporated within – not replaced by – the new system.

The violence of the late 1830s, as we have seen, left a lasting impression on the Huddersfield Union authorities. The rhetoric of the anti-Poor Law movement framed the Guardians' deliberations for decades. Much to the frustration of the central authorities, every attempt to intervene in local affairs met with resistance, if not from the Guardians themselves, then from township authorities; in the words of one despairing inspector, 'the very absurdity of this [local] feeling makes it all the more difficult to combat'.[64] By no means all of the Guardians were prepared to exploit the rhetoric of the anti-Poor Law years, yet even the advocates of new workhouse construction were careful to distinguish their arguments from those of the central authorities. It was the condition of children, the sick and the insane which they emphasised, not the need for a workhouse test. Even so, the campaign for workhouse construction was not to bear fruit for a generation.

# 9

# The workhouse system from a local perspective

The Union workhouse was intended to be the centrepiece of Poor Law policy after 1834. Its very layout and appearance were designed to make a powerful impression on the poor. In southern England, as we have seen, workhouse construction proceeded apace during the late 1830s. In the industrial North, however, the progress of workhouse construction was slow and uneven. In industrial Lancashire and the West Riding of Yorkshire, for example, only a few workhouses were authorised to be built before 1848. As Figure 5.5 shows, most of the Unions in this area which did embark on workhouse building during this period were located in rural districts (Halifax and Chorlton being the main exceptions). Many Boards of Guardians in the heart of the industrial North, especially those most associated with anti-Poor Law protest, refused to build new workhouses until the 1860s. Unions like Todmorden, Huddersfield and Rochdale continued to run a number of township poorhouses instead of building a single Union workhouse. This chapter examines the evolution of workhouse policy in the Huddersfield Union, whose Guardians were particularly hostile to the policies of the central authorities.

## The survival of local institutions

The idea of the workhouse was not in itself new to the industrial North. In the Huddersfield area, for example, many institutions bearing this name were established during the late eighteenth and early nineteenth centuries.[1] Most of them were small, undistinguished buildings, housing up to twenty inmates at most, set aside for those paupers who could not be provided for in any other way. Only Huddersfield itself boasted a workhouse to compare with those of other medium-sized towns, accommodating in 1835 about one hundred inmates. While it was a less substantial institution than those of larger cities like Manchester or Leeds,[2] it dwarfed most of the others in the future Huddersfield Union. Work was provided for some paupers in most of these township institutions; yet, as elsewhere in the industrial North, the term 'poorhouse' was often preferred to 'workhouse'.[3] There was certainly little resemblance between these institutions and the model workhouses recommended by the Poor Law Commissioners in 1835. In general, the old poorhouses catered for a far smaller population, and were frequently adapted from existing buildings. The logic of the purpose-designed Union institution was foreign to all but the largest cities of the industrial North.

The question of workhouse provision was first addressed by the Huddersfield Board of Guardians during 1839, after the townships had relinquished responsibility for the provision of out-door relief. There was little local support for the idea of building a new Union workhouse; instead, five of the local township poorhouses (at Huddersfield, Almondbury, Golcar, Honley and Kirkheaton) were adopted for the use of the Union. In 1843, the Guardians took formal responsibility for a sixth institution – a new vagrant office at Huddersfield. All these institutions remained in use for at least twenty years, so that in 1847 the Huddersfield Union (together with nearby Rochdale) maintained more workhouses than any other Union in the country.[4] Far from being an endorsement of the new workhouse system, local practice flew in the face of central policy, which was to concentrate large numbers of inmates in a single institution. The very word 'Union', popularly adopted as a shorthand for the reformed workhouse, marked an essential distinction between institutions under the old law and the new.

Although the central authorities intended all workhouses to be managed as Union institutions, subject to uniform regulations, many Northern poorhouses retained some if not all of the characteristics of township institutions. Those in Huddersfield were small, even by the standards of neighbouring Unions such as Rochdale and Dewsbury. During the early 1840s, workhouse repairs continued to be financed by individual townships, rather than (as the new law required) from the common fund of the Union. In the case of the new vagrant office, the master's salary was (illegally) charged to local funds, on the grounds that the institution benefited the township alone. It was only in 1847, following action by the persistent district auditor, that the Guardians fell into line.[5] The day-to-day management of workhouses was left in the hands of poorly paid masters and matrons, under the intermittent supervision of local Guardians. On several occasions during the 1840s, local workhouse officers were discovered to be acting almost independently, lodging their relatives in workhouses, embezzling supplies, taking unauthorised leave and sending children out as apprentices without official consent.[6] While masters and matrons in many other Unions found loopholes in official regulations, in Huddersfield their autonomy was almost taken for granted. The required workhouse records were rarely kept, and official surveillance was intermittent. In the course of an official investigation during 1847–8, it emerged that there had been a total breakdown of communication

Figure 9.1 Huddersfield Vagrant Office, c. 1869 (*Source*: Ramsden Collection, KDA)

between the matron of the Huddersfield workhouse, the Board of Guardians and their workhouse medical officer.[7]

The continuation of a large number of small workhouses had important implications for the kind of disciplinary regime that could be imposed. In 1848, the Huddersfield workhouse itself was certified as capable of holding 100 inmates, with the four others accommodating a total of 130 inmates. Figure 9.1 (which shows the vagrant office in 1869, shortly before its closure) indicates the small scale and quasi-domestic character of the Union's most punitive institution. The workhouses themselves were far from the imposing, disciplinary institutions envisaged in 1834; one official dismissed them as 'nondescript houses'.[8] Dispersed among existing communities, they signally failed to sequester their inmates in 'well-regulated' spaces. The walls of township institutions remained permeable, their inmates relatively accessible. As one resident of Slaithwaite, writing of the 1860s, recalls:

> One thing I could not understand then or now: it was why an idiot whom we only knew as 'Yellow John', hailing from some workhouse in Golcar, should have had the freedom of visiting our village at pleasure . . . Such a menace would not be tolerated for an hour nowadays.[9]

At Huddersfield, as in other Unions where the Guardians refused to build a new workhouse, the central authorities pressed for improvements in classification and discipline. The small township poorhouses were supposed to be

Table 9.1. *Composition of the workhouse population in the Huddersfield Union, 1841–1861*

| Workhouse | Children under 14 M | Children under 14 F | Adults under 60 M | Adults under 60 F | Adults over 60 M | Adults over 60 F | Total |
|---|---|---|---|---|---|---|---|
| *1841* | | | | | | | |
| Huddersfield | 15 | 22 | 14 | 20 | 17 | 7 | 95 |
| Almondbury | 12 | 9 | 9 | 16 | 6 | 3 | 55 |
| Golcar | 4 | 4 | 9 | 5 | 5 | 4 | 31 |
| Honley | 7 | 6 | 2 | 12 | 6 | 6 | 39 |
| Kirkheaton | 2 | 1 | 1 | 5 | 4 | 3 | 16 |
| Total | 40 | 42 | 35 | 58 | 38 | 23 | 236 |
| *1851* | | | | | | | |
| Huddersfield | 15 | 5 | 30 | 25 | 18 | 10 | 103 |
| Almondbury | 4 | 3 | 10 | 12 | 8 | 4 | 41 |
| Golcar | 1 | 0 | 5 | 5 | 8 | 2 | 21 |
| Honley | 4 | 1 | 3 | 5 | 5 | 5 | 23 |
| Kirkheaton | 6 | 4 | 9 | 14 | 10 | 4 | 47 |
| Total | 30 | 13 | 57 | 61 | 49 | 25 | 235 |
| *1861* | | | | | | | |
| Huddersfield | 1 | 0 | 39 | 35 | 32 | 7 | 114 |
| Almondbury | 1 | 1 | 10 | 12 | 10 | 4 | 38 |
| Golcar | 2 | 0 | 7 | 4 | 8 | 2 | 23 |
| Honley | 1 | 1 | 2 | 7 | 3 | 3 | 17 |
| Kirkheaton | 21 | 14 | 2 | 11 | 3 | 0 | 51 |
| Total | 26 | 16 | 60 | 69 | 56 | 16 | 243 |

*Source:* Census Enumerators Books

treated as if they were separate wards in a single building, enabling classification between (rather than within) institutions. In 1840, the Huddersfield Guardians agreed to concentrate their able-bodied male paupers at Golcar, and able-bodied females at Honley.[10] The available evidence suggests that this policy had little effect before the 1850s. In 1853, following renewed official pressure, the Guardians decided 'as far as practicable' to implement their initial classification, and also to send children to Kirkheaton and the elderly to Huddersfield.[11] Judging from the census data shown in Table 9.1, this new policy did have some effect. Until the 1850s, however, the workhouses appear to have been treated as local institutions, serving primarily

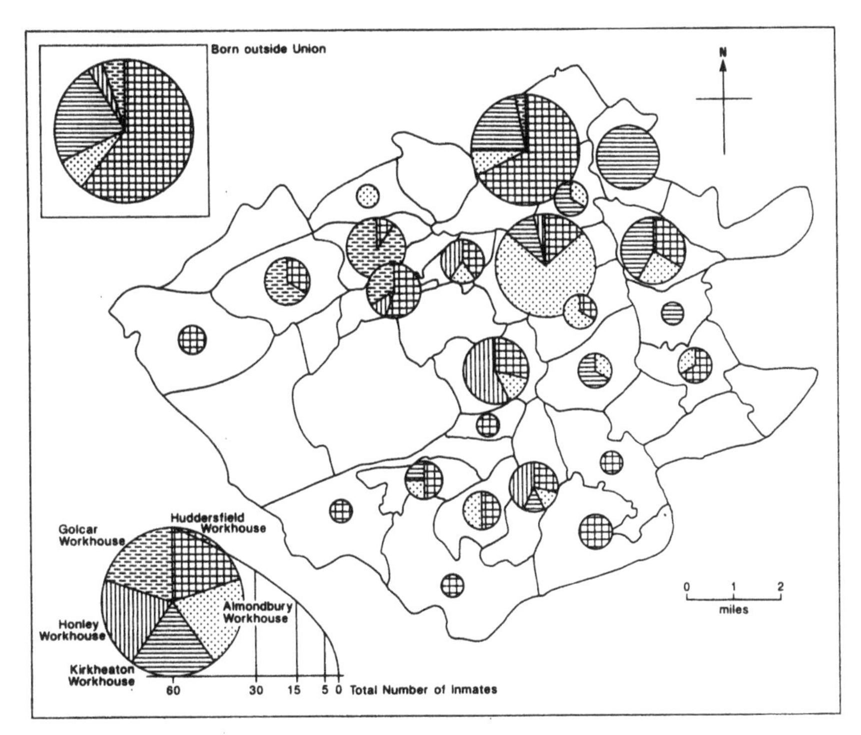


Figure 9.2 Birthplaces of Huddersfield Union workhouse inmates, 1851 (*Source*: Census. See Figure 8.2 for base map).

township rather than Union needs. Figure 9.2 shows the birthplaces of the inmates of the five workhouses on census night in 1851. The fact that each workhouse drew upon a distinct local catchment suggests that location (or perhaps settlement) was an important consideration in the allocation of paupers to institutions. Such patterns were not unique to Huddersfield; elsewhere in the industrial North, prior to the construction of large new Union workhouses, township institutions were often used in similar ways.[12] After 1851, the proportion of inmates who were born locally (as recorded in the census) fell significantly, indicating a move towards a genuine Union workhouse policy. Nevertheless, the policy of classification between workhouses was to remain incomplete, as each workhouse continued to house a wide variety of paupers.

## The coming of the Union workhouse

The campaign for a new workhouse in the Huddersfield Union was a protracted one. Throughout the 1840s and 1850s, central inspectors argued that the retention of several small, dispersed and unspecialised institutions undermined both classification and control.[13] Amongst the Guardians, however, there was initially little support for a new building, and the issue was frequently overshadowed by disputes over the division of the Union (see pp. 137–8). Overcrowding in existing workhouses periodically led the Board to consider providing additional accommodation, although proposals for a single Union-wide institution received little support. A particularly dramatic crisis in institutional provision occurred during 1847–8, when local workhouses (already under pressure during a period of economic distress) were swelled with typhus cases. The use of a temporary fever hospital and new sick wards drew unprecedented public attention to the condition of the Huddersfield workhouse. Lurid reports appeared in the press, playing on the moral and physical consequences of overcrowding in a workhouse where children associated freely with adults, and the sick were mixed with the healthy.[14] The crisis occasioned by the typhus outbreak thus precipitated a broader debate, placing the question of workhouse policy firmly on the agenda of local politics[15] (see below).

Although the crisis of 1847–8 in some ways strengthened the case for new workhouse construction, it also highlighted a recurrent dilemma facing the central authorities. For the Guardians' response – the improvement of existing accommodation – effectively undermined proposals for an entirely new workhouse. Subsequent measures, such as the adoption of a stricter classification policy in 1853, had the same effect. Thus a proposal to build a new workhouse in February 1855 was to be heavily defeated.[16] Local hostility to new capital expenditure remained strong, despite the increasing pressure on existing institutions. In 1857, local reformers finally succeeded in passing a motion for a new workhouse in the Holmfirth district; yet in the Guardians' elections of the following year their opponents swept them from power.[17] In July 1858, the official Poor Law inspector reported to London that 'the present Board of Guardians are endeavouring to give out-door relief as much as possible, particularly in sick and infectious cases, to make it appear that the existing workhouses are sufficient'.[18]

The debates on the Huddersfield Board during 1857–8 suggested that an influential section of local opinion now favoured the construction of at least one new workhouse, although few Guardians approved of the idea of a single Union institution. Having reached an *impasse* with the hostile Board elected in 1858, the central authority (urged on by a new generation of inspectors) took what turned out to be a decisive step, by refusing to sanction further expenditure on existing workhouses and threatening to close several of them by Order.[19] This fiscal weapon, also deployed in other Northern Unions during this period, appears to have achieved some success. In 1859, following renewed controversy over the treatment of insane inmates (discussed below), the Huddersfield Guardians finally agreed to provide a

Figure 9.3 The Crosland Moor Workhouse: main block

new workhouse at Deanhouse, Honley, and to close three of their existing institutions. The central authorities sanctioned the proposal, although they considered the building (designed for 200 inmates) inadequate; as one inspector wearily observed in February 1860, 'There will now be three workhouses instead of five, which is something gained'.[20] But even this was not to be achieved, as one of the three institutions scheduled for closure remained in use. And soon the new workhouse was itself subject to official criticism, another inspector complaining (just six years later) that it was 'much to be regretted that so ill-arranged and incomplete a building was ever erected'.[21]

The local workhouse debate was prolonged rather than resolved by the opening of the Deanhouse institution. As well as being too small for local needs, it was comparatively inaccessible to the town of Huddersfield. (One inspector was subsequently to describe the site as 'a bleak hill in the coldest and most snowy part of England'[22].) Year after year, the pressure for new accommodation grew. After a series of false starts, the Guardians finally decided on a site for a new workhouse at Crosland Moor, near Huddersfield. The cost of building the new workhouse, initially designed to house 413 inmates, was estimated at £21,000, though the final figure was nearer £25,000. Its opening (in August 1872) was followed by the closure of the old township workhouses, leaving the Union with just two institutions.[23] (The

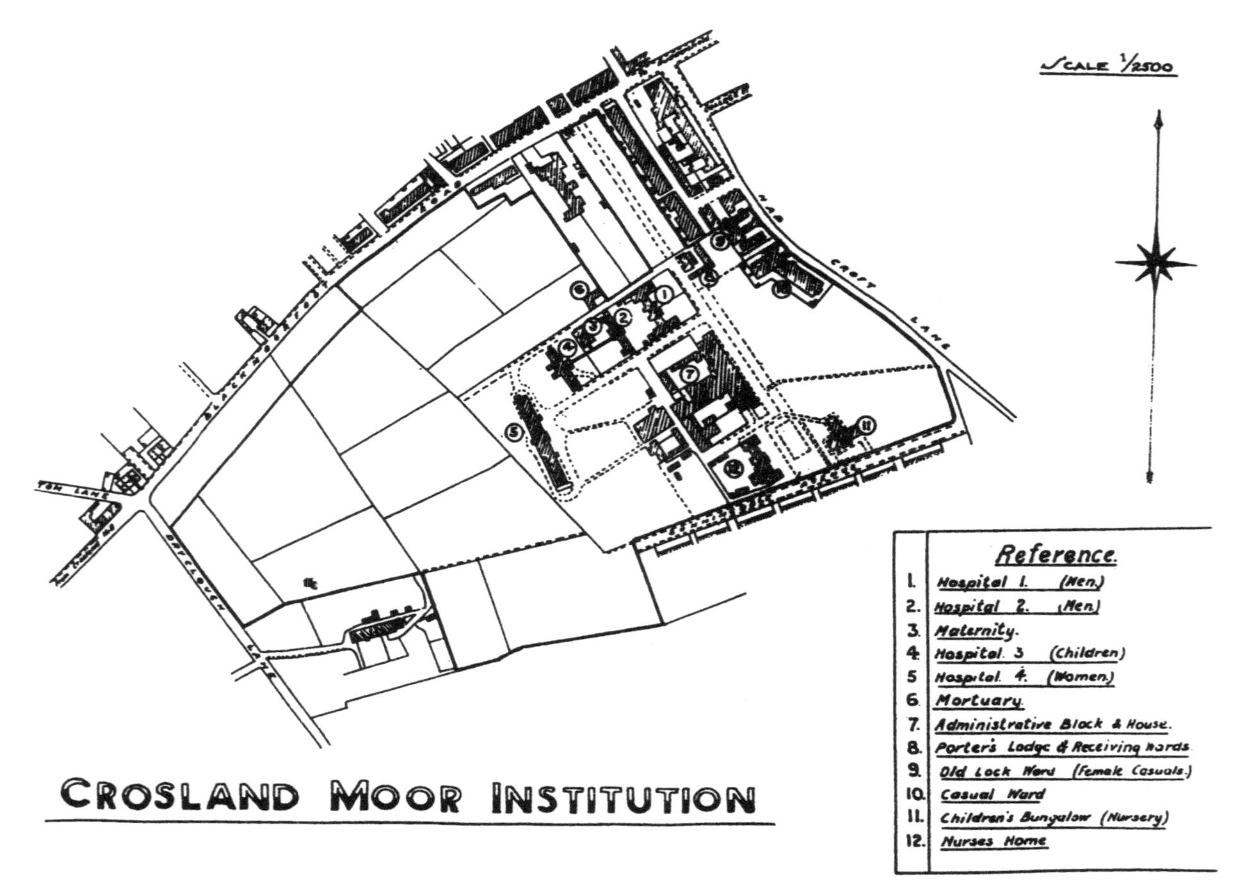


Figure 9.4 The Crosland Moor Institution, 1932 (*Source*: MH 66/681)

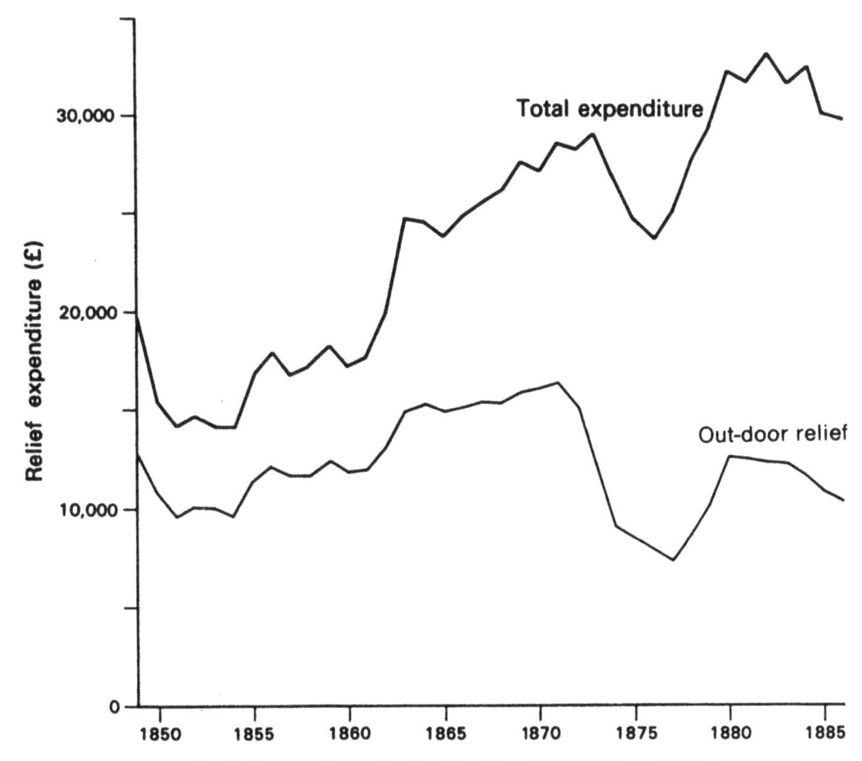


Figure 9.5 Poor Relief expenditure in the Huddersfield Union, 1849–1886 (*Source*: Annual Poor Rate Returns)

addition of new vagrant wards at Crosland Moor also enabled the Guardians to close the Huddersfield vagrant office.) 1872 thus marked the real end of the era of township poorhouses, and the beginning of the *Union* workhouse system proper. The main building at Crosland Moor, 'larger in dimensions than any mansion for miles around'[24] according to one triumphal account, housed over 200 paupers (Figure 9.3). Adjoining it, there was a substantial infirmary, infectious wards, schools and vagrant wards. Figure 9.4 shows the site as it was in 1932, by which time an infants' home (no. 11) and another large hospital ward (no. 5) had been added, and the schools converted into nurses' homes (no. 12). Crosland Moor housed a complex of institutions, staffed by a veritable regiment of nurses, attendants, porters and industrial trainers.

The building of the Crosland Moor institution allowed a significant net increase in local workhouse accommodation. In 1885, for example, the total workhouse capacity of the Union was put at 774 inmates,[25] more than three times the limit certified for the five workhouses maintained in 1848. This increase in capacity more than matched the rate of population growth over the same period. Moreover, indoor pauperism rose from around 7% of total pauperism immediately before the mid 1860s to 19% during the early 1880s (see Figure 8.4). Although out-door paupers still heavily outnumbered indoor (by four to one), institutional expenditure accounted for a growing proportion of local relief costs. As Figure 9.5 shows, out-door relief became progressively less significant, relative to other items of expenditure, in the years after 1862. Increasing in importance were indoor relief, workhouse loan repayments, asylum bills and officers' salaries. Asylum charges grew particularly rapidly; after 1870, indeed, they constituted over 10% of annual expenditure. Together with indoor relief and loan repayments, they exceeded out-door relief expenditure in every single year between 1873 and 1886, suggesting that, in financial terms, the local relief system had moved decisively towards institutional provision. The 1870s marked a significant discontinuity in relief expenditure in England and Wales as a whole;[26] the trend appeared more dramatic in Huddersfield because the Union had been so slow to fall in line with official policy.

## Local controversies over classification: children and the insane

The issue of classification loomed large in local debates over workhouse provision during the 1850s and 1860s, as it did in the country at large (chapter 4). How far the small, unsegregated workhouses of the Union could cater for the particular needs of specific groups of paupers was a question keenly debated by the Guardians. After 1847, central inspection of local workhouses was augmented by two new bodies – the workhouse schools inspectorate and the visiting Commissioners in Lunacy – whose regular reports maintained the pressure of central surveillance on local officials. Yet it is evident from debates over the treatment of pauper children and the insane that interpretations of classification frequently diverged (as seen in chapter 6). The following account traces the course and impact of these debates from a predominantly local perspective.

### *Workhouse children*

Between 1858 and 1884, between a fifth and a quarter of all the workhouse inmates in the Huddersfield Union were under the age of sixteen (Figure 9.6), a proportion somewhat lower than the national average. Most of these children either were orphans or had been abandoned by their parents, although a fluctuating proportion of them entered the workhouse with their relatives. Whilst children constituted a minority of the workhouse population, they received increasing attention in the course of local debates over classification.

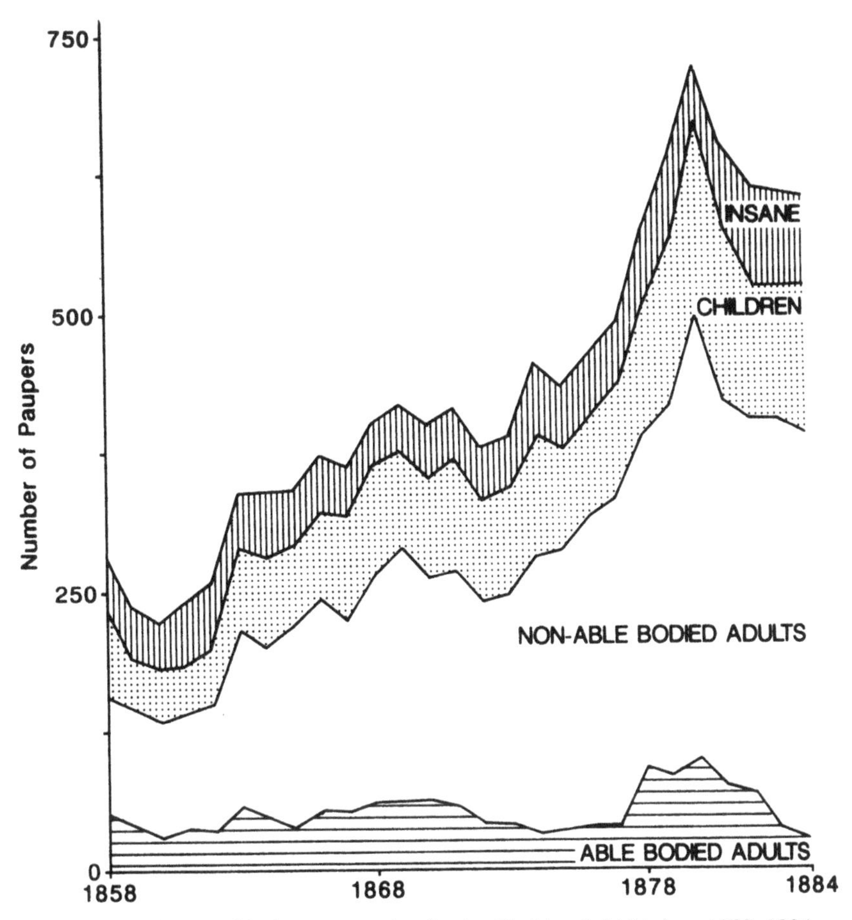


Figure 9.6 Classes of indoor pauperism in the Huddersfield Union, 1858–1884 (*Source*: Annual Returns of Pauperism)

The condition of workhouse children was placed firmly on the agenda of local policy-makers after the first visit of a workhouse schools inspector, in 1847. 'In Huddersfield workhouse', the inspector reported, 'the teacher is a pauper who cannot spell and is obviously incompetent . . . Two children only could read the Testament. None could repeat the first Commandment. None knew the Queen's name or that of the County'.[27] Such criticism gained local currency during 1848, when the whole question of workhouse provision entered the public arena (see p. 152). A campaign conducted in the columns of the *Leeds Mercury* drew particular attention to the state of the over-crowded children's wards at Huddersfield. One report drew a contrast between the local workhouse and the brand new moral and industrial school at Leeds (Figure 6.1). 'We are entitled to ask', it was concluded, 'whether Huddersfield and Leeds are in the same England'.[28] A similar stance was taken by the workhouse schools inspectorate, which continued to stress the educational and moral advantages of specialist institutions for pauper children. During the late 1840s, however, the Poor Law Board refused to support proposals for a separate school in Huddersfield, on the grounds that such expenditure would ruin any prospect of an entirely new workhouse.[29]

Although there were some changes in local provision for workhouse children after 1847, including an extension of the practice of sending children to schools outside the workhouse, official criticism continued unabated. The Poor Law Board itself pressed for further improvements in classification. After 1853, the children of the Union were concentrated at the Kirkheaton workhouse (described by one inspector as a 'respectable looking cottage').[30] Even this was only partially implemented, for a dozen adult paupers remained at Kirkheaton, and children continued to be admitted to the other workhouses. During the debates of the late 1850s (discussed above), Poor Law inspectors renewed their criticism of the moral 'contamination' of workhouse children through constant association with adult inmates.[31] The increasing pressure on existing accommodation during the 1860s inevitably resulted in more children being housed in the general workhouses. At Kirkheaton, meanwhile, standards of education and supervision continued to be criticised. Shortly before the closure of the workhouse in 1872, the local medical officer described its medical facilities as a 'sham'.[32]

At the new Crosland Moor institution, the children were housed in a large separate building at some distance from the main workhouse block (no. 12 on Figure 9.4). In addition to a salaried schoolmistress, the Guardians soon appointed an industrial trainer for the boys (in 1873), another for the girls (1880) and a bandmaster (1889). In justifying these appointments, Union officials were not slow to exploit official rhetoric; the Guardians' clerk, for example, emphasised the need to 'stamp out and eradicate from the children ... every atom of pauper taint'.[33] Nevertheless, local standards of education, care and training continued to be questioned in London. Following a series of particularly critical official reports on the workhouse school, the Guardians decided (in 1878) to try sending the children out to neighbouring Board Schools during the day. Reporting on the positive effects of this experiment, the clerk cited improvements 'not only in [the children's] appearance and health, but in education, behaviour and conduct generally'.[34] The practice of sending workhouse children out to regular schools became increasingly popular; by 1894, it had been adopted in nearly two-thirds of all Unions.[35]

Debates over local provision for workhouse children under the new Poor Law were frequently framed in the language of classification. During the 1840s, official reports were dominated by the problem of association between adults and children; during the 1850s and 1860s, schools inspectors continued to condemn the absence of trained teaching staff and specialist treatment. The concentration of the children at the new workhouse site in 1872, which was more to the liking of Poor Law inspectors than their educational colleagues, once more revived the issue of their 'contamination' by contact with adult paupers. It was this concern with the 'pauper taint' which supplied the rationale for the Guardians' decision to send the children out to school, and which eventually provided the impetus for the local adoption of the boarding-out system (in 1889) and the building of entirely separate children's homes (in 1901 and 1904). Significantly, the moving force behind both these schemes was Emily Siddon, an active campaigner for female suffrage and the first woman elected to the Huddersfield Board. Women's prominence in national-level pressure groups for the reform of pauper training (discussed in chapter 6) reflected their growing impact at a local level in Unions like Huddersfield.[36]

*The pauper insane*

The absolute number of local workhouse inmates classified as insane grew slowly during the mid nineteenth century; as a proportion of the total, there was a slight decline after the mid 1860s (Figure 9.6). Most of them were certified as chronic cases, either 'idiots' or 'imbeciles'; of the 52 insane inmates recorded in January 1857, for example, the vast majority were described as 'idiotic', most of these 'from birth'.[37] These paupers were amongst the most dependent of all inmates, and were more likely to spend years rather than weeks or months inside the workhouse.[38] As Table 9.2 shows, the Union maintained an increasing number of insane paupers in asylums, especially after the mid 1850s; this was reflected in the growing proportion of local relief expenditure accounted for by asylum bills.[39] Nevertheless, workhouses continued to accommodate a large number of the insane throughout the period; indeed, in 1881 the proportion of the chargeable insane relieved in local workhouses (36%) was greater than it had been in 1844 (28%). The growing costs of both asylum maintenance and workhouse provision for the insane thus provided a powerful stimulus to local debate.

It was not until the 1850s that the treatment of the insane became an important local issue in its own right. Before then, the problem was subsumed within more general debates over workhouse management. On their first visit to Huddersfield (in March 1849), the Commissioners in Lunacy complained that there were no special arrangements for insane paupers, who were mixed indiscriminately with other inmates.[40] Local Guardians were said to prefer keeping insane paupers in the workhouse because of the cost of asylum bills. Frustration at their lack of influence with the Poor Law authorities led the Commissioners to step up their attack. In the late 1850s, a

Table 9.2. *Insane paupers chargeable to the Huddersfield Union, 1844–1881*

| | Asylums | Hospitals | Workhouses | Out-door | Total |
|---|---|---|---|---|---|
| 1844 | 31 | 1 | 18 | 15 | 65 |
| 1851 | 32 | 0 | 46 | 24 | 102 |
| 1861 | 70 | 0 | 57 | 18 | 145 |
| 1871 | 133 | 1 | 50 | 20 | 204 |
| 1881 | 162 | 3 | 93 | 2 | 260 |

*Sources:* Returns in PP 1845 XXVIII; PP 1883 LVIII; MH 12/15072; MH 12/15078

new Commissioner visited the workhouses of the Huddersfield Union, including – for the first time – those at Almondbury, Golcar and Honley. Classification was, he said, non-existent, the wards were overcrowded and there was inadequate ventilation. His conclusion was far-reaching:

> A fundamental reformation is needed, embracing a due regard to the necessities of those who are not only helpless, but who are incapable of expressing their wants or of making just complaints of the treatment to which they have been subjected.[41]

The local response to these criticisms was generally defensive, although the Guardians did make some minor improvements.[42] A minority of Guardians argued for the construction of entirely new separate wards for the insane. Yet even this scheme was not wholeheartedly approved by the central authorities. While the Poor Law Board feared that it would divert plans for an entirely new workhouse, the Lunacy Commissioners warned that standards of treatment in workhouse lunatic wards could never match those of asylums.[43] The local support for these proposals reflected a rather different set of priorities to those of the central authorities. Firstly, the notion of classification was interpreted less as a technique of appropriate treatment than as a way of enhancing workhouse discipline in general. It was in this vein that one local campaigner spoke of the workhouses as 'nurseries of insanity and crime, dens of moral degradation and infamy', while another stressed the consequences of forcing 'both young and old, sick and well, into constant contact with jabbering idiotcy [*sic*]'.[44] The overriding concern here was with the mental and moral impact of association between the sane and insane; a 'more powerful means of extending idiotcy' one critic argued, 'could scarcely be devised'.[45] Another important local stimulus to reform was provided by the increasing cost of asylum maintenance. If a large proportion of asylum patients could be removed to secure workhouse accommodation, as had been the case at nearby Oldham and Halifax, substantial savings could be made. By 1867, nearly half of the Northern industrial Unions provided at least one separate workhouse ward for lunatics; several of these housed over 100 inmates[46] (see above, pp. 109–10).

The proposal to provide new lunatic wards at Huddersfield was stalled following the Guardians' elections of March 1858 (see above). In the longer term, however, the treatment of the insane remained an important consideration in local debates over workhouse provision. The decision to provide a new workhouse at Deanhouse, for example, was influenced by adverse publicity over the suicide of a Honley workhouse inmate (described as suffering from 'temporary insanity'),[47] as well as by the growing notoriety of the Huddersfield Union in the wake of the Lunacy Commissioners' public attack on Poor Law provision in 1859 (see chapter 6). Above all else, however, it was the escalating cost of asylum maintenance which kept the care of lunatics at the centre of local debate. In January 1862, for example, the weekly cost of maintenance at the Wakefield asylum was said to be more than twice the marginal cost of retaining a pauper in the workhouse. There was thus a considerable financial incentive for the Guardians to transfer any suitable cases to the workhouse, and deputations were regularly sent to the asylum for precisely this purpose.[48] In theory, any non-violent cases were liable to be transferred to the workhouse. One of the pauper lunatics supported by the Huddersfield Guardians at Wakefield during this period was Charlotte Phillips, a local schoolteacher and the estranged wife of the Yorkshire educationalist, George 'January Searle' Phillips. Her long spell in the asylum rather than the workhouse may have been justified on medical grounds, but other considerations could conceivably have played their part.[49]

Although the new workhouses at Deanhouse and Crosland Moor enabled an expansion of local provision for the chronic insane, the lack of specialist supervision continued to be a major source of official complaint. Until 1892, when all such cases were transferred to Deanhouse (under the care of special attendants),[50] they remained the responsibility of the general workhouse staff and other inmates. The Guardians appear to have seriously considered the appointment of specialist staff on one previous occasion, in 1873, when they approved a plan to build a special lunatic ward at Crosland Moor. Once again, this proposal was provoked by the rising cost of asylum maintenance and pressure on existing accommodation. The plan was dropped following the opening of the Wadsley asylum, near Sheffield, and the provision of a new government grant for the care of pauper lunatics in asylums.[51]

## Conclusion: local institutional landscapes

This chapter has provided a local perspective on the evolution of workhouse policy in the Huddersfield Union. As Figure 8.3 suggests, the process of policy-making involved the representatives of many different agencies and authorities. Particular attention has been paid here to the awkward relations between the Huddersfield Guardians and a variety of central inspectorates. Rather less has been said about the relationships within and between the local authorities themselves (a primary concern of chapter 8). A more comprehensive study of local institutional provision would need to consider the role of institutions beyond the domain of the Poor Law. The Huddersfield Dispensary, founded in 1814, for example, remained a major source of support for the sick poor throughout the period. The associated Infirmary, established in 1831, provided facilities for the treatment of acute illness and accident (though not of chronic or infectious disease).[52] These institutions helped to shape the local role of the workhouse, 'from the outside', as it were. The Guardians of the Huddersfield Union took out a regular subscription to the Infirmary, as they did to other specialist institutions in the region. At the same time, many other local authorities came to provide alternative institutional provision for some categories of potential or actual paupers. The foundation of a large Model Lodging House by the Huddersfield Improvement Commissioners (in 1854), the establishment of a temporary fever hospital by the Huddersfield Corporation (in 1873), and the creation of local Board Schools (following the 1870 Education Act) had important implications for the local role of the workhouse system.

The well-defined boundary around the Poor Law which the 1834 reformers aimed to construct – through the 'workhouse test' – was effectively reinforced by the decisions of a range of local administrators beyond the Poor Law. The Model Lodging House, for example, was specifically designed to provide an alternative to both the workhouses and local lodging houses frequented by the casual poor, a morally elevating kind of accommodation for the independent labourer. Those who were unable to afford its charges were turned away and referred to the Poor Law authorities.[53] Similarly, the temporary fever hospital established by the Borough in 1873 (on the very site of the old workhouse) was restricted to non-pauper patients only. In justifying this policy, the local Medical Officer of Health remarked that 'the honest, though poor working man, holds the pauper in great contempt and considers it a stigma upon his character to be associated with him'.[54] These views were not, of course, unique to Huddersfield; an official report in 1882 advised that the successful conversion of workhouse facilities into isolation hospitals depended on 'the extent to which the buildings could be severed in the eyes of the general public from their former pauper connexions'.[55] Neither were such views confined to hospital administrators. In 1878, for example, the Huddersfield School Board attempted (unsuccessfully) to prevent workhouse children from attending its schools. One local newspaper commented that the Board 'could not have been more particular if every workhouse child necessarily had the measles'.[56]

The evolution of workhouse policy and practice in Huddersfield thus reflected not only the relationships between central and local authorities, but also the shifting pattern of debate and discussion at the local level. Frequently, the suggestions of central inspectors were undermined or simply ignored; even the hardening attitude of the central authorities in the late 1850s did not secure immediate success. Only when official criticism found a substantial local echo did the Guardians commit themselves to significant reform. The local press, in particular, appears to have played an increasingly important role in campaigns over workhouse policy, not least because of the influence of individual Guardians with particular newspapers and their editors. These local campaigns were not simply miniature reflections of broader debates. The principle of classification, as we have seen, was subject to different interpretations in different times and places. It is one of the tasks of the historical geographer to map the complex relationships which shaped these conflicting interpretations of social policy.

# Conclusion

> When [workhouse inmates] are spoken to, when they are set to work, where they dress, when they eat, when they retire to rest, when husbands and wives, and children and parents, meet, when sickness falls, when the old die and the young are born, when the 'House' is entered and when it is left – it matters not when or where – the statement is ever proclaimed, 'You are a pauper', and you ought to know it.[1]

In June 1876, Robert Barnes, a fifty-year-old man officially described as an 'itinerant conjurer', was killed in an accident at a workhouse near Huddersfield. The event gave rise to a routine report from the workhouse medical officer to the Local Government Board in London, stating the circumstances in which the accident had occurred. According to the doctor's report, Robert Barnes was a 'dissipated man' who spent his late years 'frequenting taverns and spending nearly all the money that came to hand in drink'. 'By his death', it was noted, 'both the police and the Poor Law officials of the town will be freed from an almost constant applicant for assistance'.[2] An inquest was considered unnecessary, for the report itself sufficed. The death of Robert Barnes was an event quickly forgotten in the history of the new Poor Law. Yet it provides a fitting conclusion to this study, highlighting some broader truths about the workhouse system. It occasioned an official document, the medical officer's report, which was dispatched to an office in London and subjected there to all the usual bureaucratic rituals. The document traced just one of the innumerable encounters between the cumbersome machinery of the new Poor Law and its object, pauperism – capricious, mobile and insubordinate pauperism – 'order's itinerant nightmare'.[3] As we have seen, the report says more about the conduct of Robert Barnes than about the event which caused his death. This moralising tone served an administrative purpose. In its evident concern with the costs of improvidence, the medical officer's report rehearsed official preoccupations with pauperism as something to be *governed.*

I have suggested in this book that the history of the new Poor Law is to a large degree a history of power relations – the power inscribed in administrative texts, the calculating power of institutional design, sometimes the power of popular resistance, frequently the power of local recalcitrance and, for the most part, the powerlessness of paupers themselves. In this context, the programme of workhouse policy outlined in 1834 has been portrayed as the product of two distinct sets of strategies; those of modern government and those of institutional discipline. The novelty of 1834 lay not in the association of poor relief provision with power in general, but rather in the particular way these strategies coalesced to form the workhouse system. The 1834 reform thus provided a radically new framework for the management of pauperism, redrawing the map of Poor Law government. However, it could not anticipate the ways in which official policy would subsequently develop. The history of workhouse policy is also, therefore, a history of struggles – struggles within the central Poor Law authority itself, between the Poor Law authorities and other sources of authority and expertise, between popular movements and the state, and between central and local authorities. Poor Law historians have focussed overwhelmingly on the last of these, narrowing the frame of their inquiries so much that other struggles and controversies have frequently been neglected. In this book, central policy has been portrayed as the provisional outcome of processes of negotiation and conflict which stretch well beyond the domain of the workhouse system itself.

Although the workhouse system was often represented as a monolithic structure, it was in practice internally differentiated; it had an historical geography. Throughout this book, I have argued the case for a geographical perspective on social policy; one which emphasises the fractured nature of nineteenth-century government and the uneven impact of surveillance. The analysis of patterns of relief regulation and workhouse construction presented here offers a genuinely national map of the workhouse system, in place of the dualisms characteristic of conventional histories of 'central' policy and 'local' practice. Such maps provide the sense of a broader context which is essential for local studies. In the latter part of the book, I examined the post-1834 workhouse system from such a local perspective; not just any local perspective, as the Huddersfield experience was clearly a particular one in several respects. Yet it is clear that, even in Huddersfield, the discourses and practices of local policy reflected far more than merely local priorities. After all, the story of Robert Barnes, the itinerant conjurer from Huddersfield, would not have survived had it not been relayed to London by a workhouse medical officer. It becomes intelligible only when situated within the wider apparatus of administrative power and official knowledge that constituted the workhouse system. It represents the moralising gaze of authority, inscribed in a fragment from the history of social policy; one more collision between power and pauperism.

# Notes

## Introduction

1 Anon., 'The English bastile', *Social Science Review* 3 (1865), 193.
2 This was the spelling most commonly used in Victorian literature on the workhouse. It is retained throughout this book.
3 Anon., 'The English bastile', 195–7.
4 Peter Bussey, cited in D. Ashforth, 'The urban Poor Law' in D. Fraser (ed.), *The new Poor Law in the nineteenth century* (London, 1976), 129.
5 M. Crowther, *The workhouse system, 1834–1929* (London, 1981), 268.
6 'It will be one of the enigmas destined to puzzle posterity, that England, which undertook to clothe, conquer and evangelize the world, would yet be baffled by its own paupers': J. Hole, *Lectures on social science and the organisation of labor* (London, 1851), 38.
7 Reports on Vagrancy: PP 1866 XXXV, 62–7.
8 J. Mozley to LGB, 29 August 1882: MH 12/15252.
9 Anon., 'The English Bastile', 197.
10 Register of Local Poor Law Officers, Hitchin: MH 9/8.

## 1 Policing society: government, discipline and social policy

1 D. Fraser, *The evolution of the British welfare state* (second edn, London, 1984), xxii–xxxi. See also M. Bulmer, J. Lewis and D. Piachaud (eds.), *The goals of social policy* (London, 1989).
2 J. Goldthorpe, 'The development of social policy in England, 1800–1914', *Transactions, Fifth World Congress of Sociology* 4 (1962), 41–56; P. Corrigan, 'State formation and moral regulation in nineteenth-century Britain' (Ph.D. thesis, University of Durham, 1977); D. Thomson, 'Welfare and the historians' in L. Bonfield, R. Smith and K. Wrightson (eds.), *The world we have gained* (Oxford, 1986), 355–78.
3 F. Piven and R. Cloward, *Regulating the poor* (New York, 1971); N. Ginsburg, *Class, capital and social policy* (London, 1979); I. Gough, *The political economy of the welfare state* (London, 1979).
4 I. Hacking, 'How should we do the history of statistics?' in G. Burchell,

C. Gordon and P. Miller (eds.), *The Foucault effect: studies in governmentality* (London, 1991), 181–95.

5 A. Silver, 'The demand for order in civil society' in D. Bordua (ed.), *The police* (New York, 1967), 6.

6 G. Rosen, 'Cameralism and the concept of medical police' in G. Rosen (ed.), *From medical police to social medicine* (New York, 1974), 142–58.

7 K. Dyson, *The state tradition in Western Europe* (Oxford, 1980), 18–21; M. Raeff, 'The well-ordered police state', *American Historical Review* 80 (1975), 1221–43.

8 M. Foucault, 'Governmentality', *Ideology and Consciousness* 6 (1979), 9–21, and *The history of sexuality, volume I* (London, 1979); P. Pasquino, 'Theatrum politicum', *Ideology and Consciousness* 4 (1978), 41–54.

9 A. Heidenheimer, 'Politics, policy and *policey* as concepts in English and continental languages', *Review of Politics* 48 (1986), 3–30; Dyson, *The state tradition*, 36–44.

10 L. Hume, *Bentham and bureaucracy* (Cambridge, 1981), 33–6; J. Annette, 'Bentham's fear of hobgoblins' in B. Fine *et al.* (eds.), *Capitalism and the rule of law* (London, 1979), 70–1; Silver, 'The demand for order in civil society'; A. Donajgrodski, 'Social police and the bureaucratic elite' in A. Donajgrodski (ed.), *Social control in nineteenth-century Britain* (London, 1976), 51–76; D. Philips, 'A new engine of power and authority' in G. Parker, B. Lenman and V. Gatrell (eds.), *Crime and the law* (London, 1980), 155–89; V. Gatrell, 'Crime, authority and the policeman-state' in F. Thompson (ed.), *The Cambridge social history of Britain, 1750–1950, volume III* (Cambridge, 1990), 243–310.

11 M. Dean, *The constitution of poverty* (London, 1991), 62, 195–6.

12 T. Malthus, *Essay on population*, 6th edn (London, 1826), 339. This was not an argument for anarchy; it suggested a form of police that followed rather than resisted the 'laws of nature'.

13 M. Foucault, 'Space, knowledge and power', *Skyline*, March 1982, 16.

14 M. Mann, 'The autonomous power of the state', *Archives of European Sociology* 25 (1984), 185.

15 *Ibid.*, 200.

16 A. Giddens, *A contemporary critique of historical materialism*, vol. II: *The nation-state and violence* (Cambridge, 1985).

17 Mann, 'The autonomous power of the state', 208. See also P. Alliès, *L'invention du territoire* (Grenoble, 1980); C. Raffestin, *Pour une géographie du pouvoir* (Paris, 1980).

18 For more general discussions, see C. Harris, 'Power, modernity and historical geography', *Annals, Association of American Geographers* 81 (1991), 671–83; J. Robinson, 'A perfect system of control? State power and "native locations" in South Africa', *Society and Space* 8 (1990), 135–62; M. Mann, *The sources of social power* (Cambridge, 1986); J. Hall (ed.), *States in history* (Oxford, 1986).

19 Silver, 'The demand for order in civil society'; Philips, 'A new engine of power and authority'; F. Mather, *Public order in the age of the Chartists* (Manchester, 1959).

20 M. Ogborn, 'Local power and state regulation in nineteenth-century Britain', *Transactions, Institute of British Geographers* 17 (1992), 215–26.

21 R. Paddison, *The fragmented state* (Oxford, 1983), 5.

22 Giddens, *A contemporary critique*; C. Dandeker, *Surveillance, power and modernity* (Cambridge, 1990); R. Sack, *Human territoriality* (Cambridge, 1986).
23 N. Rose, 'Calculable minds and manageable individuals', *History of the Human Sciences* 1 (1988), 187.
24 J. Bowring (ed.), *The works of Jeremy Bentham* (London, 1843), vol. IV, 39.
25 F. Mort, *Dangerous sexualities: medico-moral politics in England since 1830* (London, 1987), 37. See also N. Rose, *The psychological complex: psychology, politics and society in England, 1869–1939* (London, 1985), 11–38; Hacking, 'statistics'; P. Corrigan and D. Sayer, *The great arch* (Oxford, 1985); P. Corrigan, 'On moral regulation' in P. Corrigan, *Social forms/human capacities* (London, 1990), 102–29.
26 M. Wiener, *Reconstructing the criminal* (Cambridge, 1990), 152.
27 S. Collini, *Public moralists: political life and intellectual thought in Britain, 1850–1930* (Oxford, 1991), 62–7.
28 M. Cullen, *The statistical movement in early Victorian Britain* (Brighton, 1975), 65–74; F. Driver, 'Moral geographies: social science and the urban environment in mid-nineteenth century England', *Transactions, Institute of British Geographers* 13 (1988), 275–87.
29 J. Prichard, *A treatise on insanity* (London, 1835), cited in Wiener, *Reconstructing the criminal*, 166. See also E. Carlson and N. Dain, 'The meaning of moral insanity', *Bulletin of the History of Medicine* 36 (1962), 130–40; M. Donnelly, *Managing the mind* (London, 1983), 135–9.
30 M. Foucault, *Madness and civilisation* (London, 1967), 241–55; Rose, *The psychological complex*, 25–9; R. Porter, *Mind-forg'd manacles* (London, 1987), 222–8; A. Scull, 'Moral treatment reconsidered' in A. Scull (ed.), *Social order/mental disorder* (Berkeley, 1989), 80–94.
31 It was in this sense that the 1834 Poor Law Report described the labouring classes in the south of England as 'demoralised': K. Snell, *Annals of the labouring poor* (Cambridge, 1985), 170–4.
32 A. Scull, 'Moral architecture: the Victorian lunatic asylum' in *Social order/mental disorder*, 213–38; R. Evans, *The fabrication of virtue* (Cambridge, 1982); F. Driver, 'Discipline without frontiers?', *Journal of Historical Sociology* 3 (1990), 272–93; M. Ogborn, 'A lynx-eyed and iron-handed system: the state regulation of prostitution in nineteenth-century Britain' (unpublished conference paper, 1989); C. Rosenberg, 'Florence Nightingale on contagion: the hospital as moral universe' in C. Rosenberg (ed.), *Healing and history* (New York, 1979), 116–36; T. Markus, 'The school as machine' in T. Markus (ed.), *Order in space and society* (Edinburgh, 1982), 201–56. See also D. Rothman, *The discovery of the asylum* (Toronto, 1971); A. King, *Buildings and society* (London, 1980); Donnelly, *Managing the mind*; J. Thompson and G. Goldin, *The hospital* (New Haven, 1975); M. Foucault *et al.*, *Les machines à guérir* (Brussels, 1976).
33 M. Foucault, *Discipline and punish* (Harmondsworth, 1977), 219.
34 R. Evans, 'Bentham's Panopticon', *Architectural Association Quarterly*, July 1971, 35. On architecture as a 'moral science', see also Rothman, *The discovery of the asylum*, 83–4.
35 F. Driver, 'Power, space and the body', *Society and Space* 3 (1985), 425–46; M. Perrot (ed.), *L'impossible prison* (Paris, 1980); C. Philo, 'Enough to drive one

mad: the organisation of space in nineteenth-century lunatic asylums' in J. Wolch and M. Dear (eds.), *The power of geography* (London, 1989), 258–90.

36 Foucault, *Discipline and punish*, 139.

37 Giddens, *A contemporary critique*, 184–6.

38 Goffman describes some of these strategies in explicitly spatial terms: see his discussion of 'free places' and 'personal territories' in 'The underlife of a public institution' in *Asylums* (Harmondsworth, 1968), 203–22.

39 C. Gordon, 'Afterword' in C. Gordon (ed.), *Power/Knowledge* (Brighton, 1980), 246.

40 Foucault, *Discipline and punish*, 205.

41 M. Foucault, 'La poussière et le nuage' in Perrot (ed.), *L'impossible prison*, 35, 49.

42 S. Tuke, 'Practical hints on the construction of pauper lunatic asylums' in *Plans of the pauper lunatic asylum at Wakefield* (York, 1819), 8.

43 *Transactions, NAPSS* (1857), xxiv. Frederic Demetz, founder of the famous Mettray reformatory colony, once described the world as a 'vast house of correction': F. D. Hill, *Children of the State*, second edn (London, 1889), 273.

44 S. Daniels and S. Seymour, 'Landscape design and the idea of improvement, 1730–1900' in R. Dodgshon and R. Butlin (eds.), *An historical geography of England and Wales* (London, 1990), 487–520.

45 J. Schmiechen, 'The Victorians, the historians and the idea of modernism', *American Historical Review* 93 (1988), 302–14.

46 H. Austin, Report on Plans of the Proposed Lodging House at Huddersfield, 17 May 1853, MH 13/95. See also chapter 9.

47 Daniels and Seymour, 'Landscape design'.

48 Associationism proposed that 'complex mental phenomena are formed from simple elements derived ultimately from sensations [and] that the mechanism by which these are formed depends on the similarity and/or repeated juxtaposition of the simple elements in space and time': R. Young, 'Association of ideas' in P. Wiener (ed.), *Dictionary of the history of ideas* (London, 1973), 111.

49 Driver, 'Power, space and the body', 427.

50 M. Ignatieff, 'State, civil society and total institutions', *Crime and Justice* 3 (1981), 171, 175–6; M. Ignatieff, *A just measure of pain* (London, 1978), chapter 3.

51 Dean, *Constitution of poverty*, 216, 190–2.

52 D. Melossi, 'Some observations on the recent literature' in D. Melossi and M. Pavarini, *The prison and the factory* (London, 1981), 196.

53 See also Burchell, Gordon and Miller, *The Foucault effect*; S. Cohen, *Visions of social control* (Cambridge, 1985); F. Driver, 'Bodies in space' in C. Jones and R. Porter (ed.), *Reassessing Foucault* (London, 1993); M. Hewitt, 'Social policy and the politics of life', *Occasional Paper, School of Social Sciences, Hatfield Polytechnic* 7 (1982).

54 M. Dear and J. Wolch, *Landscapes of despair* (Cambridge, 1987); G. Kearns and C. Withers (eds.), *Urbanising Britain* (Cambridge, 1991); King, *Buildings and society*; Ogborn, 'Local power and state regulation'; F. Paterson, *Out of place: public policy and the emergence of truancy* (London, 1989); C. Philo, 'Fit localities for an asylum', *Journal of Historical Geography* 13 (1987), 398–415; C. Smith, *Public problems: the management of urban distress* (New York, 1988); D. Ward, *Poverty, ethnicity and the American city, 1840–1925* (Cambridge, 1989).

55 B. Luckin, 'Towards a social history of institutionalization', *Social History* 8

(1983), 87–94; M. Finnane, 'Asylums, families and the state', *History Workshop Journal* 20 (1985), 134–48.

56 Ignatieff's contention that Foucault has a 'state-centred conception of social order' is thus wide of the mark ('State, civil society and total institutions', 184).

57 R. Johnson, 'Educating the educators: "experts" and the state, 1833–1839' in A. Donajgrodski (ed.), *Social control in nineteenth-century Britain* (London, 1977), 79. The 'revolution in government' is discussed in chapter 2.

58 S. Cohen, 'The punitive city: notes on the dispersal of social control', *Contemporary Crises* 3 (1979), 340. See also J. Lowman, 'Conceptual issues in the geography of crime: toward a geography of social control', *Annals, Association of American Geographers* 76 (1986), 81–94.

59 Melossi, 'Some observations', 195.

## 2 Social policy, liberalism and the mid-Victorian state

1 F. Poynter, *Society and pauperism* (London, 1969), xxii.

2 E. Chadwick, 'The new Poor Law', *Edinburgh Review* 63 (1836), 490.

3 LGB, Reports on the Poor Law in Foreign Countries, PP 1875 LXV, 5–6.

4 G. Simmel, 'The poor', *Social Problems* 13 (1965–6), 122.

5 R. Dworkin, *Taking rights seriously* (London, 1977), 184; E. Halévy, *The growth of philosophic radicalism* (London, 1928), 175–6.

6 P. Corrigan and D. Sayer, *The great arch* (Oxford, 1985), 125–8; M. Wiener, *Reconstructing the criminal* (Cambridge, 1990), 141–56; M. Ignatieff, *A just measure of pain* (London, 1978), 211–15.

7 F. Bédarida, 'L'Angleterre Victorienne: paradigme du *laissez-faire*', *Revue Historique* 261 (1979), 79–98.

8 A. Briggs, 'The welfare state in historical perspective', *Archives of European Sociology* 2 (1961), 222; D. Wolfe, 'Mercantilism, Liberalism and Keynesianism', *Canadian Journal of Social and Political Theory* 5 (1981), 69–96.

9 M. Mann, *The sources of social power* (Cambridge, 1986), chapter 15; J. Brewer, *The sinews of power* (London, 1989).

10 Corrigan and Sayer, *The great arch*, 118. See also T. Novak, *Poverty and the state* (Milton Keynes, 1988), 52.

11 K. Polanyi, *The great transformation* (New York, 1945), 77–102. Also, K. Polanyi, 'Our obsolete market mentality' in G. Dalton (ed.), *Primitive, archaic and modern economies* (New York, 1968), 59–77; M. Dean, *The constitution of poverty* (London, 1991), 156–72.

12 A. Dicey, *Lectures on the relationship between law and public opinion in England during the nineteenth century* (London, 1905).

13 H. Perkin, 'Individualism versus collectivism in nineteenth-century Britain: a false antithesis', *Journal of British Studies* 17 (1977), 105–18. See also G. Kearns, 'Private property and public health reform in England, 1830–1870', *Social Science and Medicine* 26 (1988), 187–99.

14 Halévy, *Philosophic radicalism*, 405.

15 L. Hume, *Bentham and bureaucracy* (Cambridge, 1981).

16 Halévy, *Philosophic radicalism*, 432. The parallels with Foucault's account of Bentham's Panopticon scheme are compelling: see chapter 1.

17 H. Parris, 'The nineteenth-century revolution in government', *Historical Journal* 3 (1960), 19–20; L. Hume, 'Jeremy Bentham and the nineteenth-century revolution in government', *Historical Journal* 10 (1967), 371–3.
18 J. Viner, 'Adam Smith and *laissez-faire*' in *The long view and the short* (Glencoe, Illinois, 1958), 213–451; A. Coats, 'Introduction' to *The classical economists and economic policy* (London, 1971), 14–17; D. O'Brien, *The classical economists* (Oxford, 1975), 272–7.
19 J. McCulloch, cited in L. Robbins, *The theory of economic policy in English classical political economy* (London, 1952), 43.
20 N. Senior, cited in M. Bowley, *Nassau Senior and classical economics* (London, 1937), 265.
21 For the parallel case of public health, see Kearns, 'Private property'
22 R. Grew, 'The nineteenth century European state' in C. Bright and S. Harding (eds.), *State-making and social movements* (Ann Arbor, 1984), 102–9.
23 Polanyi, *The great transformation*, 140.
24 *The 1834 Poor Law Report*, ed. S. Checkland and E. Checkland (Harmondsworth, 1974), 9.
25 D. Ashforth, 'The Poor Law in Bradford, 1834–1871' (unpublished Ph.D. thesis, Bradford, 1979), 29.
26 S. Finer, *The life and times of Sir Edwin Chadwick* (London, 1952), 42.
27 PLC, *Third Annual Report* (1837), 5.
28 'Essay on the Poor Laws' (1796): Bentham Papers, CLIIIa, 73 (my emphasis). See also Dean, *The constitution of poverty*, 137–55.
29 *1834 Poor Law Report*, 127; Chadwick, 'The new Poor Law', 501.
30 J. Marshall, *The Old Poor Law, 1795–1834* (London, 1968), 13–14.
31 N. Senior, *Three lectures on the rate of wages* (London, 1830), ix–x.
32 *Ibid.*, x.
33 *1834 Poor Law Report* (1834), 132; Senior, *Three Lectures*, ix.
34 *1834 Poor Law Report* (1834), 127.
35 *Ibid.*, 147.
36 J. Wade, *Appendix to the Black Book* (London, 1835), 38–9.
37 M. Blaug, 'The myth of the old Poor Law', *Journal of Economic History* 23 (1963), 151–84. For a commentary on subsequent work, see G. Boyer, *An economic history of the English Poor Law, 1750–1850* (Cambridge, 1990).
38 K. Williams, *From pauperism to poverty* (London, 1981), 55.
39 N. Senior, 'English Poor Laws' in *Historical and philosophical essays* (London, 1865), vol. II, 97.
40 P. Dunkley, *The crisis of the old Poor Law, 1795–1834* (New York, 1982), iv.
41 A. Brundage, *The making of the new Poor Law, 1832–1839* (London, 1978), 15. See also E. Hobsbawm and G. Rudé, *Captain Swing* (New York, 1968); Dunkley, *The crisis of the old Poor Law*, 255–67.
42 Dunkley, *The crisis of the old Poor Law*, 173.
43 H. Parris, 'The revolution in government'; cf. R. Cosgrove, *The rule of law: A. V. Dicey, Victorian jurist* (London, 1980).
44 O. Macdonagh, 'The nineteenth-century revolution in government', *Historical Journal* 1 (1958), 52–67.
45 R. MacLeod (ed.), *Government and expertise* (Cambridge, 1988), 5–8.

46 J. Hart, 'Nineteenth-century social reform: a Tory interpretation of history', *Past and Present* 31 (1965), 39–61; S. Finer, 'The transmission of Benthamite ideas, 1820–1850', in G. Sutherland (ed.), *Studies in the growth of nineteenth-century government* (London, 1972), 11–32.
47 Hart, 'Nineteenth-century social reform'; P. Richards, 'State formation and class struggle, 1832–1848' in P. Corrigan (ed.), *Capitalism, state formation and Marxist theory* (London, 1980), 49–52. Macdonagh subsequently revised his initial formulations in *Early Victorian government, 1830–1870* (London, 1977), esp. 3–8.
48 U. Henriques, 'Jeremy Bentham and the machinery of social reform' in H. Hearder and H. Loyn (eds.), *British government and administration* (Cardiff, 1974), 169–86.
49 H. Clokie and J. Robinson, *Royal Commissions of Inquiry* (Stanford University, 1937), 75–9.
50 Finer, 'The transmission of Benthamite ideas', 11–32; Henriques, 'Jeremy Bentham', 172.
51 J. Austin, 'Centralisation', *Edinburgh Review* 85 (1847), 232–3.
52 E. Hennock, 'Central/local relations in England', *Urban History Yearbook* (1982), 40.
53 D. Marshall, 'The role of the Justice of the Peace' in Hearder and Loyn (eds.), *British government and administration*, 155–68.
54 H. Finer, in J. Harris, *British government inspection* (London, 1955), viii.
55 G. Kitson Clark, 'Statesmen in disguise', *Historical Journal* 2 (1959), 19–39.
56 Corrigan and Sayer, *The great arch*, 125.
57 P. Bartrip, 'British government inspection, 1832–1875', *Historical Journal* 25 (1982), 605–26; Bartrip, 'State intervention in nineteenth-century Britain', *Journal of British Studies* 23 (1983), 63–83.
58 MacLeod, *Government and expertise*, 9.
59 C. Bellamy, *Administering central–local relations, 1871–1919* (Manchester, 1988), chapter 4.
60 W. Carson, 'Symbolic and instrumental dimensions of early factory legislation' in R. Hood (ed.), *Crime, criminology and public policy* (London, 1974), 107–38.
61 P. Richards, 'The state and early industrial capitalism', *Past and Present* 83 (1979), 91. Compare D. Roberts, *Victorian origins of the British welfare state* (New Haven, 1960), 34.
62 P. Bartrip, 'Success or failure? The prosecution of the early Factory Acts', *Economic History Review* 38 (1985), 436.
63 PLC Minutes, 4 November 1834: MH 1/1.
64 W. Greenleaf, 'Toulmin Smith and the British political tradition', *Public Administration* 53 (1975), 25–55.
65 R. Lambert, *Sir John Simon, 1816–1904, and English social administration* (London, 1963).
66 V. Gatrell, 'Crime, authority and the policeman-state' in F. Thompson (ed.), *The Cambridge social history of Britain, 1750–1950* (Cambridge, 1990), 261–2; Bellamy, *Administering central–local relations*.
67 H. Preston-Thomas, *The work and play of a government inspector* (Edinburgh, 1909), 363–4.

**3 Rational landscapes? The geography of Poor Law government**

1 G. Simmel, 'The poor' (1908), in *Social Problems* 13 (1965–6), 129.
2 G. Young, *Victorian England: portrait of an age* (London, 1936), 40–1.
3 A. de Tocqueville, *Correspondence and conversations with Nassau Senior* (London, 1872), vol. I, 203–4.
4 J. Wade, *Appendix to the Black Book* (London, 1835), 45; Anon., 'Second Session of the reformed House', *Westminster Review* 21 (1834), 431.
5 P. Mandler, 'The making of the new Poor Law *redivivus*', *Past and Present* 117 (1987), 132–3.
6 A. Brundage, 'The landed interest and the new Poor Law', *English Historical Review* 87 (1972), 27–48; 'The new Poor Law and the cohesion of agricultural society', *Agricultural History* 47 (1974), 405–17.
7 This is the approach in K. Williams, *From pauperism to poverty* (London, 1981). Williams' work is discussed below, and in chapter 5.
8 A. Digby, 'The principles of 1834', *Middlesbrough Centre, Occasional Paper* 1 (1985), 1–14.
9 E. Chadwick, 'The new Poor Law', *Edinburgh Review* 63 (1836), 533.
10 PLB, *Fourteenth Annual Report* (1848), Appendices A1 and A4.
11 M. Wiener, *Reconstructing the criminal* (Cambridge, 1990), 103–9.
12 PLC to Normanby, 31 July 1840, HO 73/56. Also, R.C. Poor Laws, PP 1909 XXXVII, 987; Chadwick, 'The new Poor Law', 527.
13 PLC, *Seventh Annual Report* (1841), 55; J. Harris, *British Government Inspection* (London, 1955), 12–21.
14 D. Roberts, *Victorian origins of the welfare state* (New Haven, 1960), 328–9.
15 Anon., 'Patronage of Commissions', *Westminster Review* 46 (1847), 229.
16 C. Mott to PLC, 2 February 1843, MH 32/57; C. Mott to E. Chadwick, 24 April 1847 (Chadwick Papers, no. 1449); W. Day to PLC, 22 January 1844, MH 32/14.
17 J. Simon to G. Sclater Booth, 10 November 1875: MH 78/135.
18 C. Bellamy, *Administering central–local relations, 1871–1919* (Manchester, 1988), 111–65.
19 F. Mouat, Memorandum, 20 June 1876: MH 32/51.
20 R.C. Poor Laws, PP 1909 XXXVII, 982.
21 PLB, *Twelfth Annual Report* (1860), 24–6.
22 Bellamy, *Administering central–local relations*, 174–5.
23 C. Bahmueller, *The national charity company* (Berkeley, 1981), 8.
24 From 1662, ecclesiastical parishes could be divided into 'townships' for Poor Law purposes. In this chapter, the term 'parish' denotes all local units of administration.
25 A 'settlement' in a particular place entitled individuals to claim for poor relief from the rates raised in that place; without a settlement, they were liable, under certain circumstances, to removal. The basic criteria of settlement were altered during the 1840s, although Union-level chargeability was not achieved until 1865: M. Caplan, 'The new Poor Law and the struggle for Union chargeability', *International Review of Social History* 23 (1978), 267–300.
26 J. Redlich and F. Hirst, *Local government in England* (London, 1903), vol. I, 106, and vol. II, 204; B. Webb and S. Webb, *English Poor Law History*, Part II

(London, 1929), vol. I, 82; H. Peake, 'Geographical aspects of administrative areas', *Geography* 15 (1930), 531–46; V. Lipman, *Local government areas, 1834–1945* (Oxford, 1949), 44; P. Perry, *Geography of nineteenth-century Britain* (London, 1975), 39.
27 G. Nicholls, *A History of the English Poor Law* (London, 1898), vol. II, 291.
28 Chadwick, 'The new Poor Law', 509–11; R.C. Poor Laws, Appendix C, PP 1834 XXXVII, 109–12.
29 *1834 Poor Law Report*, ed. S. Checkland and E. Checkland (Harmondsworth, 1974), 155; PLC, *First Annual Report* (1835), 10–2. See also P. Dunkley and W. Apfel, 'English rural society and the new Poor Law', *Social History* 10 (1985), 58.
30 PLC, *First Annual Report* (1835), 12.
31 Brundage, 'The landed interest', 30–42.
32 P. Dunkley, 'The landed interest and the making of the new Poor Law', *English Historical Review* 88 (1973), 840–1; A. Digby, *Pauper palaces* (London, 1978), 55–61; M. Rose, 'Poor Law administration in the West Riding of Yorkshire' (unpublished D. Phil. thesis, University of Oxford, 1965), 91–111; E. Midwinter, *Social administration in Lancashire, 1830–1860* (Manchester, 1969), 15–20.
33 A. Power, Report on Lancashire and the West Riding of Yorkshire, 25 November 1836: MH 32/63.
34 C. Mott to J. Lefevre, 15 September 1838: MH 32/57.
35 S. C. Poor Relief, PP 1844 X, Q174.
36 B. and S. Webb, *English Poor Law History*, Part II, vol. I, 116.
37 These powers were granted in the case of smaller parishes only in 1844.
38 PLC, *Ninth Annual Report* (1843), 54–71; B. Webb and S. Webb, *Statutory authorities for specific purposes* (London, 1922), 107–51.
39 Webb and Webb, *Statutory authorities*, 139–44.
40 N. Senior, *Remarks on the opposition to the Poor Law Amendment Bill* (London, 1841), 87; PLC, *Fourth Annual Report* (1838), 4–16.
41 R v. PLC (re. Whitechapel), 6 Ad & El 34; R v. PLC (re. St Pancras), 6 Ad & El 1.
42 PLC, *Seventh Annual Report* (1841), 22–3, and *Ninth Annual Report* (1843), 70–1; S.C. Poor Relief (1844), Q71.
43 T. Gilbert, *Considerations on the Bills for the better relief and employment of the poor* (London, 1787), 26.
44 B. Webb and S. Webb, *English Poor Law history*, Part I: *The old Poor Law* (London, 1927), 171n.
45 S. C. Poor Relief (1844), Q47–51. Another return lists 76 incorporations in 1834, containing 1075 parishes: PP 1844 XL.
46 Return of Gilbert's and Local Act Unions: PP 1856 XLIX.
47 H. Zouch, *Remarks upon the late resolutions of the House of Commons* (Leeds, 1776), 19–22, 41. Compare E. Martin, 'From parish to Union' in E. Martin (ed.), *Comparative development in social welfare* (London, 1972), 33; A. Coats, 'Economic thought and Poor Law policy in the eighteenth century', *Economic History Review* 13 (1960–1), 39–51.
48 Webb, *The old Poor Law*, 272–6; E. Copleston, *A second letter to the Right Honourable Robert Peel* (Oxford, 1819), 74–9; PLC, *First Annual Report* (1835),

13; PLC, *Third Annual Report* (1837), 40–3; S. C. Poor Relief (1844), Q6, 78, 100, and 141.
49 E. Duncombe, *Gilbertise the new Poor Law* (York, 1841), 26.
50 A. Power, Reports of 24 and 31 January 1837: MH 32/63.
51 R v. PLC (re Allstonefield Incorporation), 11 Ad & El 558.
52 *Hansard*, 24 May 1844: vol. 74, col. 1486.
53 Williams, *From pauperism to poverty*, 75.
54 PLC, *First Annual Report* (1835), 37.
55 PLC, *Second Annual Report* (1836), 8.
56 T. Frankland Lewis to J. Russell, 13 August 1838, Russell Papers.
57 J. Russell to PLC, 26 June 1837: HO 73/52.
58 PLC, *Fourth Annual Report* (1838), 29–30.
59 PLC, *Eighth Annual Report* (1842), 13–14.
60 PLC, *First Annual Report* (1835), 7–8.
61 E. Tufnell to E. Chadwick, 28 May 1847: HO 45/2019.
62 PLB, *Fifth Annual Report* (1853), 17–31.
63 Williams, *From pauperism to poverty*, 65.
64 E. Chadwick, 'The statutory sanction of out-door relief', MH 32/118; C. Mott to PLC, 20 July 1841, MH 32/57; U. Corbett to PLB, 10 April 1865, MH 32/13; A. Doyle to PLB, 21 April 1865, MH 32/19.
65 LGB, *Fifteenth Annual Report* (1886), xiv–xv. See M. Mackinnon, 'English Poor Law policy and the crusade against outrelief', *Journal of Economic History* 47 (1987), 603–25.
66 See, for example, J. Davy, 'Report on the Manchester out-relief regulations', 8 December 1877: MH 32/98.
67 Bellamy, *Managing central–local relations*, 234.
68 R.C. Poor Laws, PP 1909 XXXVII, 746.
69 G. Brodrick, 'Local government in England' in J. Probyn (ed.), *Local government and taxation* (London, 1875), 29.

**4 Designing the workhouse system, 1834–1884**

1 PLC Minutes, 4 November 1834: MH 1/1.
2 F. Head, cited in S. Jackman, *Galloping Head* (London, 1958), 63.
3 A. Power, Report on Stamford Rivers Incorporation, 15 April 1835, MH 32/63.
4 E. Tufnell, cited in M. Crowther, *The workhouse system* (London, 1981), 41.
5 G. Gilbert Scott, *Personal and professional recollections* (London, 1879), 77.
6 Anon., 'New Poor Law workhouses', *Illustrated London News*, 7 November 1846; Anon., 'Patronage of Commissions', *Westminster Review* 46 (1847), 227.
7 H. Maunsell, *Political medicine* (Dublin, 1839), 43–4.
8 PLC, *Fifth Annual Report* (1839), 35 and Appendix B10. The Limerick workhouse, for example, was designed for 2,500 inmates; PLC, *Thirteenth Annual Report* (1847), Appendix A14.
9 M. Titmarsh [W. Thackeray], *The Irish sketch-book* (London, 1843), vol. II, 131; S. Ferguson, 'Architecture in Ireland', *Dublin University Magazine* 29 (1847), 693–708.

10 A. Dickens, 'The architect and the workhouse', *Architectural Review* 160 (1976), 345–52; G. White (ed.), *In and out of the workhouse* (Ely, 1978), 19–32.
11 G. Wythen Baxter, *The book of the bastiles* (London, 1841).
12 A. Pugin, *Contrasts* (London, 1841).
13 C. Mott to J. Lefevre, 13 April 1839, MH 12/15065.
14 *1834 Poor Law Report*, ed. Checkland (Harmondsworth, 1974), 31.
15 Report on the Poor Law Commission: PP 1840 XVII, 30.
16 J. Bowring (ed.), *The works of Jeremy Bentham* (London, 1843), vol. VIII, 362.
17 *The Architectural Magazine* 2 (1835), 511.
18 A. Scull, 'The domestication of madness', *Medical History* 27 (1983), 248.
19 Crowther, *Workhouse system*, 42–3.
20 PLC, *Eighth Annual Report* (1842), Appendix A3.
21 R.C. Popular Education, PP 1861 XXI Pt. I, 358; Dr Rhodes, 'Classification of paupers by workhouses', *Poor Law Conferences* (1895), 266–77.
22 PLC, *Second Annual Report* (1836), 6.
23 This term is discussed in chapter 1.
24 Report on the Poor Law Commission (1840), 37.
25 E. Smith, Report on Provincial Workhouses: PP 1867–8 LX, 7.
26 See, for example, J. Locke, *Report of the Board of Trade to the Lords Justices in the year 1697* (London, 1789); Anon., *An account of several workhouses for employing and maintaining the poor* (London, 1732); F. Eden, *The state of the poor* (London, 1797), vol. I, 312–13; J. Becher, *The anti-pauper system* (London, 1828).
27 Returns in *House of Commons Reports*, vol. IX, 299–539; PP 1854–5 XLVI; J. Taylor, 'The unreformed workhouse' in E. Martin (ed.), *Comparative development in social welfare* (London, 1972), 57–64.
28 H. Stokes, 'Cambridge parish workhouses', *Proceedings of the Cambridge Antiquarian Society* 59 (1911), 131; J. Cole (ed.), *Down workhouse lane* (Littleborough, 1984), 42; J. Sykes, *Slawit in the sixties* (Huddersfield, 1926), 128.
29 R. Pashley, *Pauperism and Poor Laws* (London, 1852), 364–5.
30 F. Cobbe, 'The philosophy of the Poor Laws', *Fraser's Magazine* 70 (1864), 386. See also J. Stallard, *Workhouse hospitals* (London, 1865).
31 Anon., 'The English Bastile', *Social Science Review* 3 (1865), 195.
32 C. Herford, *The education of orphan pauper children* (Manchester, 1865), 20.
33 S. Turner, *Reformatory schools* (London, 1855), 2.
34 W. Lindsay, 'The family system as applied to the treatment of the chronic insane', *Journal of Mental Science* 16 (1871), 504. See also A. Scull, *Decarceration* (Cambridge, 1984); C. Philo, 'Enough to drive one mad: the organisation of space in nineteenth-century lunatic asylums' in J. Wolch and M. Dear (eds.), *The power of geography* (London, 1989), 258–90.
35 F. Driver, 'Discipline without frontiers? Representations of the Mettray Reformatory Colony in England, 1840–1880' *Journal of Historical Sociology* 3 (1990), 272–93; W. Parry-Jones, 'The model of the Geel lunatic colony and its influence on the nineteenth-century asylum system in Britain' in A. Scull (ed.), *Madhouse, mad-doctors and madmen* (London, 1981), 201–7; Philo, 'Enough to drive one mad'.
36 P. Hollis, *Ladies elect* (London, 1987), 284. See also J. Rendall, *The origins of modern feminism* (London, 1985), 183–6, 267–70; Anon., 'The work of women as

Poor Law Guardians', *Westminster Review* 67 (1885), 386–94; L. Twining, 'Women as official inspectors', *Nineteenth Century* 35 (1894), 489–94.
37 J. Butler (ed.), *Woman's work and woman's culture* (London, 1869), xxxvii.
38 R. Hodgkinson, *The origins of the National Health Service* (London, 1967), 495.
39 *Ibid.*, 481–2.
40 E. Smith, Report on workhouse dietaries, PP 1866 XXXV, 24–5.
41 E. Smith, Report on Metropolitan Workhouses, PP 1866 LXI, 35. See also his Report on Provincial Workhouses, PP 1867–8 LX.
42 *Hansard*, 8 February 1867: vol. 185, col. 163.
43 R.C. Poor Laws, Appendix 1, PP 1909 XXXIX, Q2383.
44 A. Brundage, *The making of the new Poor Law* (London, 1978), 54.
45 PLB, *Twenty-first Annual Report* (1869), 41–5 and Appendix A6. Also, P. Smith, *Hints and suggestions as to the planning of Poor Law buildings* (London, 1904); K. Williams, *From pauperism to poverty* (London, 1981), 116–18.
46 M. Mackinnon, 'English Poor law policy and the crusade against outrelief', *Journal of Economic History* 47 (1987), 615; M. Rose, 'The crisis of poor relief in England, 1860–1914' in W. Mommsen (ed.), *The emergence of the welfare state in Britain and Germany* (London, 1981), 59.
47 P. Wood, 'Finance and the urban Poor Law: Sunderland Union, 1836–1914' in M. Rose (ed.), *The poor and the city* (Leicester, 1985), 34–9. See also chapter 9.
48 C. Bellamy, *Administering central–local relations, 1871–1919* (Manchester, 1988), 241–2.
49 Williams, *From pauperism to poverty*, 96–100, 143–4.
50 M. Wiener, *Reconstructing the criminal* (Cambridge, 1990), 153.
51 Cobbe, 'Philosophy of the Poor Laws', 386.
52 LGB, *Fourth Annual Report* (1875), 47–9.

### 5 Building the workhouse system, 1834–1884

1 A. Digby, *Pauper palaces* (London, 1978), 232.
2 K. Williams, *From pauperism to poverty* (London, 1981).
3 *Ibid.*, 75–9, Table 4.33(a).
4 Authorisations to lease or rent property are excluded from this analysis.
5 Williams, *From pauperism to poverty*, 218–19, 224.
6 *Ibid.*, 77.
7 M. Crowther, *The workhouse system* (London, 1981), 62; Williams, *From pauperism to poverty*, 77.
8 HGM, 24 September 1841.
9 A. Dickens, 'The architect and the workhouse', *Architectural Review* 160 (1976), 345–52.
10 G. Gilbert Scott, *Personal and professional recollections* (London, 1879), 79. See also D. Cole, 'Sir George Gilbert Scott' in P. Ferriday (ed.), *Victorian architecture* (London, 1963), 178.
11 G. White, (ed.), *In and out of the workhouse* (Ely, 1978); Digby, *Pauper palaces*, 62–75.
12 R. Hodgkinson, *The origins of the National Health Service* (London, 1967), 469–87.

13 Williams, *From pauperism to poverty*, 123–5.
14 Reports on Vagrancy: PP 1866 XXXV.
15 LGB, *First Annual Report* (1872), 54–63; *Third Annual Report* (1874), xviii–xx.
16 Report of the Departmental Committee on Vagrancy, PP 1906 CIII, Appendix 13; D. Jones, *Crime, protest, community and police in nineteenth-century Britain* (London, 1982), 190–1.
17 A. Freeman, *Hints on the planning of the Poor Law buildings* (London, 1904), 13–16; C. Craven, *A night in the workhouse* (Keighley, 1887); Crowther, *The workhouse system*, 259–61.
18 S. Cox to LGB, 19 June 1876, MH 12/16222; LGB, *Seventh Annual Report* (1878), xi; *Ninth Annual Report* (1880), 205–6; P. Smith, *Hints and suggestions on the planning of Poor Law buildings* (London, 1901), 10–23.
19 There was to be a dramatic increase in workhouse expenditure during the 1890s: Williams, *From pauperism to poverty*, 224.
20 P. Wood, 'Finance and the urban Poor Law: Sunderland Union, 1836–1914', in M. Rose (ed.), *The poor and the city* (Leicester, 1985), 34–9.
21 E. Smith, Report on Metropolitan Workhouses: PP 1866 LXI, 35.
22 K. Maiwald, 'An index of building costs in the United Kingdom, 1845–1938', *Economic History Review* 7 (1954), 187–203.

**6 Classifying the poor: maps of pauper-land**

1 T. Murray-Browne, 'The education and future of workhouse children', *Poor Law Conferences* (London, 1883), 97–8.
2 For statistics, see PLB, *Eleventh Annual Report* (1858), Appendix B3; LGB, *Fifteenth Annual Report* (1886), Appendix 42.
3 PLC, *Fourth Annual Report* (1838), 140.
4 *Ibid.*, 146, 149.
5 A. Austin, Memorandum on Kay-Shuttleworth to Ebrington, 9 August 1848, MH 12/15229.
6 PLC, *Report on the training of pauper children* (London, 1841), 217.
7 *Ibid.*, 350; PLC, *Fourth Annual Report* (1850), 145.
8 PLB, *Second Annual Report* (1850), 12–16.
9 PLB, *Twelfth Annual Report* (1860), 18.
10 CCE Minutes: PP 1849 XLII, 161.
11 L. Twining, 'Workhouse education', *Transactions, NAPSS* (1861), 332.
12 H. Synnot, 'Little paupers', *Contemporary Review* 24 (1874), 970.
13 M. Carpenter, 'On the education of pauper girls', *Trans. NAPSS* (1862), 290.
14 F. D. Hill, 'The family system for workhouse children', *Contemporary Review* 15 (1870), 241; F. D. Hill, *Children of the state* (London, 1868); Anon., 'Pauper girls', *Westminster Review* 37 (1870), 463.
15 F. Driver, 'Discipline without frontiers? Representations of the Mettray reformatory, 1840–1880', *Journal of Historical Sociology* 3 (1990), 272–93.
16 S. C. Criminal and Destitute Juveniles, PP 1852 VII, 466.
17 J. Fletcher, 'Statistics of the farm school system on the continent', *Journal of the Statistical Society of London* 15 (1852), 39.
18 S. C. Criminal and Destitute Juveniles, PP 1852 VII, Q959; Anon., 'Schools of

industry', *Chambers' Miscellany*, 14 (1846); J. Leigh, 'Juvenile offenders and destitute pauper children' in Viscount Ingestre (ed.), *Meliora* vol. II (London, 1853), 81–9.

19 S. C. Criminal and Destitute Children, PP 1852 XXIII, Q2770–1; S. Turner and T. Paynter, *Report on the system and objects of La colonie agricole at Mettray* (London, 1846).

20 Fletcher, 'Statistics of the farm school system', 37.

21 CCE Minutes, PP 1852–3 LXXIX, 43; CCE Minutes, PP 1856 XLVII, 36–7; S.C. Poor Relief, PP 1862 X, Q6193.

22 S. C. Criminal and Destitute Juveniles (1852), Q2387–2524; J. Symons, 'On industrial training' in A. Hill (ed.), *Essays on educational subjects* (London, 1857), 37–56.

23 CCE Minutes, PP 1856 XLVII, 73.

24 S.C. Poor Relief, PP 1861 IX, Q12947–13089; S.C. Poor Relief, PP 1862 X, Q4241–4656, Q6516–6920.

25 Mins CCE (1856), 106–10; SC Poor Relief (1861), Q12294–5.

26 LGB, *Third Annual Report* (1874), 311–94.

27 Driver, 'Discipline without frontiers', 273, 287.

28 A. Doyle, *Proposed district school on the system of Mettray* (London, 1873), 7.

29 Report on Cottage Homes, PP 1878 LX.

30 K. Williams, *From pauperism to poverty* (London, 1981), 221; W. Chance, *Children under the Poor Law* (London, 1897), 135–67; A. Freeman, *Hints on the planning of Poor Law buildings* (London, 1904), chapter 8.

31 M. Monnington and F. Lampard, *Our Poor Law schools* (London, 1898), 14–20.

32 F. D. Hill, *Children of the state* (1889 edn), 90–1.

33 *Ibid.*, 89–93; F. D. Hill, 'The system of boarding out pauper children', *Economic Journal*, 3 (1893), 62–73.

34 S. Pinel, cited in S. Tuke, 'Practical hints on the construction and economy of pauper lunatic asylums', in *Plans of the Pauper Lunatic Asylum lately erected at Wakefield* (York, 1819).

35 S. C. Medical Relief, PP 1844 IX, Q9835.

36 E. Chadwick to R. Lutwidge, 17 and 29 May 1847, MH 19/168.

37 D. Mellet, *The prerogative of asylumdom* (New York, 1982), 140.

38 LGB, *Fifteenth Annual Report* (1886), 127.

39 PLB, *Twelfth Annual Report* (1865), 266.

40 F. Anstie, 'Insane patients in London workhouses', *Journal of Mental Science* 11 (1865), 327–36.

41 Tuke, Practical hints', 6.

42 J. Bucknill, 'The custody of the insane poor', *Asylum Journal* 25 (1858), 461. See also C. Philo, 'Fit localities for an asylum', *Journal of Historical Geography* 13 (1987), 401–4.

43 *Supplement to the Twelfth Annual Report of the Commissioners in Lunacy*, PP 1859.1 IX, 12. See also R. Lutwidge to Lord Ebrington, 16 August 1849, MH 19/168.

44 J. Forster to PLB, 11 March 1856, MH 19/168; *Supplement to the Twelfth Annual Report of the Commissioners in Lunacy*, 11; S.C. Lunatics, PP 1859 III, Q641–3, Q1573.

45 Return of workhouse lunatic wards, PP 1863 LII.

46 P. McCandless, 'Build! Build! The controversy over the care of the chronically insane in England, 1835–1870', *Bulletin for the History of Medicine* 53 (1979), 553–74.
47 G. Ayers, *England's first state hospitals* (London, 1971).
48 LGB, Circular to Poor Law Unions, 1884, MH 19/170; D. Tuke, *The past and present provision for the insane in Yorkshire* (London, 1889), 27.
49 R. Cane to PLB, 6 January 1868, MH 32/9; J. Darlington to LGB, 1 March 1879, MH 12/14744.
50 R. Cane to LGB, 24 January 1877: MH 32/10.
51 W. Lumley to R. Lutwidge, 10 July 1847, MH 19/168.
52 Dr Campbell, Report on Deanhouse Workhouse, 12 July 1873: MH 12/15087.
53 C. Robertson, 'The care and treatment of the insane poor', *Journal of Mental Science* 13 (1867), 299.

**7 The politics of territory: the anti-Poor Law movement**

1 [N. Senior], *Remarks on the opposition to the Poor Law Amendment Act, by a Guardian* (London, 1841), 70.
2 M. Rose, 'The anti-Poor Law agitation' in J. Ward (ed.), *Popular movements, 1830–1850* (London, 1970), 82; N. Edsall, *The anti-Poor Law movement, 1834–1844* (Manchester, 1971), 143; J. Knott, *Popular opposition to the 1834 Poor Law* (London, 1986); R. Wells, 'Resistance to the new Poor Law in the rural South', *Middlesbrough Centre, Occasional Paper* 1 (1985), 15–53.
3 G. Phillips, *Walks around Huddersfield* (Huddersfield, 1848), 11.
4 M. Hovell, *The Chartist movement* (London, 1925), 86.
5 D. Thompson, *The Chartists* (London, 1983), 30, 258; S. Weaver, *John Fielden and the politics of popular radicalism* (Oxford, 1989).
6 A similar point has been made about the relationship between Chartism and Owenism: see G. Claeys, *Citizens and saints: politics and anti-politics in early British socialism* (Cambridge, 1989), 223–47; J. Harrison, *Robert Owen and the Owenites in Britain and America* (London, 1969), 219–32.
7 G. Stedman Jones, 'The language of Chartism' in J. Epstein and D. Thompson (eds.), *The Chartist experience* (London, 1982), 3–58.
8 F. Driver, 'Tory Radicalism? Ideology, strategy and locality in popular politics during the 1830s', *Northern History* 27 (1991), 120–38.
9 R. Oastler, *Eight letters to the Duke of Wellington* (London, 1835), 95n.
10 R. Oastler, *The right of the poor to liberty and life* (London, 1838); C. Driver, *Tory Radical: the life of Richard Oastler* (New York, 1946).
11 *Leeds Times*, 11 February 1837; *Halifax Express*, 1 April 1837.
12 *Northern Star*, 27 January 1838.
13 M. Fletcher, *Migration of agricultural labourers* (Bury, 1837), 4.
14 Weaver, *John Fielden*, 297.
15 P. Pickering, 'Class without words', *Past and Present* 112 (1986), 144–62; Knott, *Popular opposition.*
16 F. Mather, *Public order in the age of the Chartists* (Manchester, 1959), 21. See also W. Napier, *The life and opinions of General Sir Charles James Napier* (London, 1857), vol. II, 8, 15, 58.

17 *Northern Star*, 24 February 1838 [my emphasis].
18 Anon., *Give it a fair trial* (Huddersfield, c. 1837).
19 E. Yeo, 'Culture and constraint in working-class movements, 1830–1855' in E. Yeo and S. Yeo (eds.), *Popular culture and class conflict* (Brighton, 1981), especially 155–63. Also, R. Storch, 'The plague of blue locusts', *International Review of Social History* 20 (1975), 61–90.
20 *Leeds Mercury*, 16 May 1837.
21 *Leeds Mercury*, 8 October 1836; Anon., *A report of the proceedings of a public meeting* (Oldham, 1836), 27–8.
22 Anon., *Great meeting ... of the factory children* (Leeds, 1833); R. Oastler, *Speech delivered at a public meeting held in the Market place, Huddersfield* (Leeds, 1833); M. Sadler, *Protest against the secret proceedings of the Factory Commission* (Leeds, 1833).
23 A. Power to PLC, 18 February 1837: MH 32/63.
24 *Halifax Express*, 1 April 1837.
25 Edsall, *The anti-Poor Law movement*, 89.
26 A. Power to PLC, 10 May 1837: MH 32/63.
27 Estimates in the press reports on 20 May 1837 varied between sixty and two hundred thousand.
28 A. Power to PLC, 20 May 1837, MH 32/63; *The Times*, 18 May 1837.
29 *Halifax Guardian*, 1 August 1837. For the police account, see G. Martin to Mr May, 30 July 1837, HO 61/19; for Oastler's, see R. Oastler, *A letter to the Bishop of Exeter* (Manchester, 1838).
30 C. Driver, *Tory Radical*, 356.
31 PLC Minutes, 15 June 1837: MH 1/11.
32 J. Russell to Melbourne, 13 August and 9 September 1837, Russell Papers; N. Gash, *Reaction and reconstruction in English politics, 1832–1852* (Oxford, 1965), 164.
33 J. Russell to PLC, 26 June 1837: HO 73/52.
34 A. Power to PLC, 22 October 1837: MH 32/63.
35 A. Power to PLC, 30 October and 20 November 1837, MH 12/14720; *Hansard* 39 (1837) col. 949; Weaver, *John Fielden*, 163–81.
36 A. Power to PLC, 14 April 1838, MH 32/64; Edsall, *The anti-Poor Law movement*, 141–3.
37 S. Morehouse to PLC, 27 April 1838: MH 12/15064. See also *Northern Star*, 7 April 1838: *Halifax Guardian*, 10 April 1838; Edsall, *The anti-Poor Law movement*, 143.
38 *Northern Star*, 21 and 28 April 1838.
39 C. Mott to F. Lewis, 6 November 1838: MH 12/15064.
40 F. Driver, 'Tory Radicalism?'.
41 *The New Moral World*, 17 November 1838.
42 H. Vincent to J. Minnikin, 26 August 1838, Vincent–Minikin Papers.
43 G. Tinker to PLC, 17 February 1838: MH 12/15064.
44 *Leeds Mercury*, 6 May 1837.
45 D. Cannadine, *Land and landlords: the aristocracy and the towns, 1774–1967* (Leicester, 1980), 42; J. Springett, 'Landowners and urban development', *Journal of Historical Geography* 8 (1982), 129–44.

46 *Huddersfield Township Meetings*, 2 April 1835, 25 March 1836 and 25 March 1837.

47 J. Garrard, *Leadership and power in Victorian industrial towns* (Manchester, 1983), 115; J. Foster, *Class struggle and the industrial revolution* (London, 1974), 52.

48 *Halifax Guardian*, 21 January 1837. This slogan refers to the separation of husbands and wives in the new workhouses.

49 *Leeds Times*, 11 February, 25 March and 8 April 1837; *Leeds Intelligencer*, 4 February and 4 March 1837.

50 A. Power to PLC, 5 April 1837: MH 12/15063.

51 Edsall, *The anti-Poor Law movement*, 89; M. Rose, 'The new Poor Law in an industrial area' in R. Hartwell (ed.), *The industrial revolution* (Oxford, 1970), 137–8.

52 F. Driver, 'Tory Radicalism?'. See also, J. Epstein, *The lion of freedom: Feargus O'Connor and the Chartist movement, 1832–1842* (London, 1983), 99; Weaver, *John Fielden*, 294–6.

53 *Leeds Intelligencer*, 6 May 1837; *Leeds Mercury*, 6 May 1837. An examination of the extensive Ellice family papers revealed no mention of Edward Ellice Jr's candidature at Huddersfield, suggesting a lack of close personal ties.

54 *Leeds Mercury*, 5 August 1837.

55 *Leeds Mercury*, 6 May 1837.

56 This account is based on numerous press reports and correspondence in the Poor Law Commissioners' files.

57 R. Russell (ed.), *The early correspondence of Lord John Russell, 1805–1840* (London, 1913), 143.

58 J. Foster, *Pedigrees of the county families of Yorkshire* (London, 1874); West Riding poll book for 1837; *Halifax Guardian*, 24 April 1838; *Halifax Express*, 10 June 1838; B. I'Anson, *The history of the Armytage or Armitage family* (London, c. 1915), 39.

59 A. Peacock, 'The Justices of the Peace and the prosecution of the Factory Acts' (unpublished Ph.D. thesis, University of York, 1982), 169.

60 J. Sutcliffe to J. Hume, 7 June 1837, Anon. to J. Russell, 8 June 1837, and W. Moore to Col. Maberly, 16 June 1837, HO 52/35; G. Tinker to PLC, 8 June 1837, and S. Morehouse to PLC, 14 June 1837, MH 12/15063.

61 Home Office to J. Armitage, 9, 10 and 21 June 1837, and Home Office to B. Batty, 21 June 1837, HO 41/13.

62 J. Russell to Harewood, 5 July 1837: HO 41/13.

63 J. Russell to Cottenham, 7 November 1837: HO 43/53. Cf. A. Power to PLC, 16 October 1837, HO 73/52.

64 J. Russell to Harewood, 24 and 26 October 1837, Harewood Papers.

65 J. Lefevre to HO, 27 October 1837: HO 73/52.

66 In a letter to Russell, Harewood also mistakenly identified John Hague as a prospective magistrate for Huddersfield. Hague subsequently served on the Dewsbury Bench: Harewood to Russell, 5 November 1837, Harewood Papers.

67 I am grateful to John Styles for clarification on this point.

68 *Leeds Mercury*, 6 May 1837; Anon., *Report of the sayings and doings at the grand yellow dinner* (Huddersfield, 1834); J. Hanson, *View extraordinary of Sir John's Huddersfield menagerie* (Leeds, 1837).

69 W. Brook to J. Russell, January 1838, HO 40/40; HGM, 29 January 1838.
70 *Northern Star*, 3 February 1838.
71 *Halifax Guardian*, 2 January 1838; *Northern Star*, 13 and 27 January 1838. On the relations between radicals and the local magistracy, see F. Driver, 'Tory Radicalism?'.
72 *Halifax Guardian*, 11 February 1837.

**8 From township to Union? The geography of Poor Law administration at a local level**

1 J. Davy, Memorandum to Mr Hibbert, 29 April 1882: MH 12/15102.
2 J. Aikin, *A description of the country from thirty to forty miles around Manchester* (London, 1795), 552. Cf. W. Crump and G. Ghorbal, *History of the Huddersfield woollen industry* (Huddersfield, 1935).
3 J. Walker, 'A sketch of the medical and general topography of Huddersfield and its adjoining district', *London Medical Repository* 10 (1818), 6.
4 S. C. Manufacturers, PP 1833 VI, Q10927–8; *Leeds Mercury*, 3 January 1827.
5 M. Marland, *Medicine and society in Wakefield and Huddersfield, 1780–1870* (Cambridge, 1987).
6 F. Driver, 'The English bastile: dimensions of the workhouse system, 1834–1884' (unpublished Ph.D. thesis, University of Cambridge, 1987), 410–13.
7 Thurstonland Town Book, Tolson Museum, Huddersfield.
8 F. Eden, *The state of the poor* (London, 1797), vol. 3, 822–5; G. Oxley, 'The permanent poor in South-West Lancashire under the old Poor Law' in J. Harris (ed.), *Liverpool and Merseyside* (Liverpool, 1969), 16–49.
9 R. C. Poor Laws, PP 1834 XXXV, 262h. See also J. Tweedy, 'The West Riding', PP 1834 XVIII, 131–3; A. Power, Report on Lancashire and Yorkshire, 25 November 1836, MH 32/63.
10 A more detailed discussion will be found in Driver, 'The English bastile', 271–9, 335–45.
11 Honley workhouse agreement, 1783 (KDA).
12 Tweedy, 'The West Riding', 808A. See also Eden, *The state of the poor*, 821 and 828; Oxley, 'The permanent poor', 35; J. Taylor, 'The unreformed workhouse, 1776–1834' in E. Martin (ed.), *Comparative development in social welfare* (London, 1972), 65.
13 In the small township of South Crosland, for example, 46 pauper children were apprenticed between 1790 and 1801; in Huddersfield itself, 353 were apprenticed between 1798 and 1844 (South Crosland Overseers' Book, KDA; Huddersfield Register of Apprentices, TM).
14 Driver, 'The English bastile', 271–3. Cf. M. Rose, 'The new Poor Law in an industrial area' in R. Hartwell (ed.), *The industrial revolution* (Oxford, 1970), 123–4.
15 PLC, *Third Annual Report* (1837), 15.
16 D. Fraser, 'The Poor Law as a political institution' in D. Fraser (ed.), *The new Poor Law in the nineteenth century* (London, 1976), 111–27; D. Fraser, 'Poor Law politics in Leeds', *Publications of the Thoresby Society* 53 (1970), 23–49; J. Garrard, *Leadership and power in Victorian industrial towns* (Manchester,

1983); D. Ashforth, 'The Poor Law in Bradford, c. 1834–1871' (unpublished Ph.D. thesis, University of Bradford, 1979), 96.

17 See Figure 8.8 in Driver, 'The English bastile', 326.

18 Ashforth, 'Poor Law in Bradford', 342. Also, Rose, 'The new Poor Law', 136–8; E. Midwinter, *Social administration in Lancashire, 1830–1860* (Manchester, 1969), 29–35.

19 *Huddersfield Township Meetings*, 1839–1840 (KDA).

20 A. Power to PLC, 22 November 1839 and 29 June 1840, MH 12/15065; A. Power to PLC, 19 January 1840, MH 32/64.

21 E. Chadwick to J. Russell, 28 March 1841, Chadwick Papers.

22 G. Robinson to PLC, 29 August 1843: MH 12/15067.

23 J. Archbold, *The Consolidated and other Orders of the Poor Law Commissioners and the Poor Law Board* (London, 1859), 253–6.

24 J. Davy, Memorandum on J. Batley to LGB, 3 July 1882, MH 12/15102. See also R. C. Poor Laws, PP 1909 XLI, Q40755–74; E. Midwinter, 'State intervention at the local level', *Historical Journal*, 10 (1967), 106–12.

25 HGM, 11 June 1841, 11 January 1850; PLC to C. Floyd, 25 June 1845, MH 12/15068.

26 HGM, 22 April 1841.

27 A. Power to PLC, 28 January and 5 February 1840; MH 12/15066.

28 According to the official record of their meetings, the Guardians seriously contemplated the division of the Union in 1840, 1841 (twice), 1842, 1844 (twice), 1849, 1855, 1858 (twice), 1865, 1871 and 1881.

29 See M. Rose, 'Poor Law administration in the West Riding of Yorkshire, 1830–1860' (unpublished D.Phil. thesis, University of Oxford, 1966), 121–2; Midwinter, *Social administration in Lancashire*, 28; Ashforth, 'Poor Law in Bradford', 119–23.

30 E. Clayton to PLB, 11 April 1867, MH 12/15083; *Huddersfield Examiner*, 17 August 1867.

31 R. Porritt to LGB, 7 July 1882; MH 12/15103. See also J. Batley to LGB, 17 March 1882, MH 12/15101; J. Batley to PLB, 7 March 1870, MH 12/15085.

32 *Huddersfield Borough Council Minutes*, 10 March and 18 April 1882 (KDA); Report on the Huddersfield Corporation Bill (1882), MH 12/15101.

33 J. Davy, Memorandum, 2 November 1883, MH 12/15105.

34 J. Davy, Memorandum on J. Heap to LGB, 10 September 1884: MH 12/15108. See also HGM, 2 May 1884; R. C. Poor Laws, PP 1909 XLI, Q40477, Q40721.

35 PLC, *Third Annual Report* (1837) 18.

36 HGM, 17 November 1848, 2 February 1855, 28 November 1879; C. Floyd to C. Clements, 2 May 1843, MH 12/15067; C. Floyd to PLB, 6 February 1854, MH 12/15073.

37 HGM, *passim*.

38 Midwinter, 'State intervention', 109–10; R. Boyson, 'The new Poor Law in North-east Lancashire', *Transactions, Lancashire and Cheshire Antiquarian Society* (1960), 37–8; P. Dunkley, 'The hungry forties and the new Poor Law', *Historical Journal* 2 (1974), 340; R. Thompson, 'The working of the Poor Law Amendment Act in Cumbria, 1836–1871', *Northern History* (1981), 133.

39 HGM, 30 September 1842.

40 C. Clements, Memorandum on C. Floyd to PLC, 10 June 1844: MH 12/15067.
41 C. Floyd to PLC, May 1842: MH 12/15066.
42 PLC, *Third Annual Report* (1837), 17; *Fourth Annual Report* (1838), 29.
43 J. Maxfield to PLC, 23 February 1844: MH 12/15067. Also, HGM, 3 February 1843.
44 A. Power to PLC, 22 February 1839: MH 32/64.
45 Even supporters of the new Poor Law disowned the workhouse test: J. Farrington, *Jim o'th Pan's journey to London, with the new Poor Law to mend, by a Collector* (Huddersfield, 1842).
46 PLC, *Eleventh Annual Report* (1845), Appendix B1.
47 S. C. Labouring Poor, PP 1843 VII, Q1361. Compare Boyson, 'The new Poor Law', 43–4; Ashforth, 'Poor Law in Bradford', 160; Rose, 'The new Poor Law', 132.
48 HGM, 25 February 1842; C. Mott to PLC, 14 April 1842, MH 12/15066.
49 C. Clements, Memorandum on C. Floyd to PLC, 7 July 1843: MH 12/15067.
50 *The Times*, 5 September 1846. Also, C. Floyd to PLC, 4 March 1847, MH 12/15069.
51 HGM, 1 October, 1852; *Huddersfield Chronicle*, 13 November 1852.
52 Boyson, 'The new Poor Law', 38–42; Rose, 'The new Poor Law', 135.
53 PLB, *Fifth Annual Report* (1853), 30–1.
54 A. Doyle, Report, 21 April 1865, MH 32/19; Corbett to PLB, 10 April 1865, MH 32/13.
55 HGM, 15 January 1858.
56 *Huddersfield and Holmfirth Examiner*, 24 February 1855.
57 *Huddersfield and Holmfirth Examiner*, 17 February 1855.
58 *Huddersfield Chronicle*, 6 March 1858.
59 *Huddersfield and Holmfirth Examiner*, 23 February 1856.
60 *Huddersfield Chronicle*, 3 April 1858.
61 J. Hall, 'The proper employment of pauper labour', *Poor Law Conferences* (1883), 519–29.
62 J. Davy to LGB, 13 February 1879: MH 32/98.
63 *Huddersfield Union: list of paupers relieved during the half-year ended Lady Day 1878*.
64 J. Davy, Memorandum to Mr Hibbert, 29 April 1882: MH 12/15102. See also J. Prest, *Liberty and locality* (Oxford, 1990), 160–5.

**9 The workhouse system from a local perspective**

1 F. Driver, 'The English bastile: dimensions of the workhouse system, 1834–1884' (unpublished Ph.D. thesis, University of Cambridge, 1987), 337–9.
2 M. Crowther, *The workhouse system* (London, 1981), 115; P. Anderson, 'The Leeds workhouse under the old Poor Law, 1726–1834', *Publications of the Thoresby Society* 56 (1980), 75–113.
3 J. Tweedy, 'The West Riding and the City of York', PP 1834 XVIII, 726.
4 PLC, *Thirteenth Annual Report* (1847), Appendix B11.
5 G. Goode, Memorandum on C. Floyd to PLC, 8 February 1844, MH 12/15067; J. Manwaring to PLC, 16 September 1846, MH 12/15068.

6 HGM, 1 May 1840, 5 February 1841, 30 September 1842, 17 July 1844, 1 September 1848.
7 A. Austin, Report on Huddersfield workhouse, 11 July 1848, MH 12/15070; T. Tatham, *Medical relief: Mr Tatham's case against the Huddersfield Guardians* (Huddersfield, 1848).
8 C. Clements, Memorandum on C. Floyd to PLC, 19 March 1845: MH 12/15068.
9 J. Sykes, *Slawit in the sixties* (Huddersfield, 1926), 128. See also M. Jagger, *The history of Honley* (Huddersfield, 1914), 158–61; J. Cole, *Down poorhouse lane: the diary of a Rochdale workhouse* (Littleborough, 1984), 49–50; R. Boyson, 'The new Poor Law in North-east Lancashire', *Transactions, Lancashire and Cheshire Antiquarian Society* 70 (1960), 50.
10 HGM, 29 May 1840.
11 C. Floyd to PLB, 16 December 1853: MH 12/15073.
12 Boyson, 'The new Poor Law', 36; E. Midwinter, *Social administration in Lancashire* (Manchester, 1969), 52–8.
13 See, for example, PLB to C. Floyd, 3 August 1848: MH 12/15070.
14 *Leeds Mercury*, 5 February, 4 March, 18 March and 6 May 1848.
15 For a more detailed discussion, see Driver, 'The English bastile', 356–64.
16 J. Cocking to H. Farnall, 23 February 1855: MH 12/15073.
17 *Huddersfield Chronicle*, 17 April 1858.
18 J. Manwaring, Memorandum, 12 July 1858: MH 12/15076.
19 J. Manwaring, Memorandum on C. Floyd to PLB, 5 October 1857, MH 12/15075; PLB to C. Floyd, 9 July 1858, MH 12/15076.
20 J. Manwaring, Memorandum on J. Cocking to PLB, 31 January 1860: MH 12/15078.
21 R. Cane, Workhouse report, 18 October 1866: MH 12/15082.
22 J. Davy, Memorandum on J. Hall to LGB, 23 May 1881, MH 12/15099. Fifty years later, Ministry of Health officials were still complaining that the institution was 'buried in the country' (Huddersfield Public Health Survey, 112: MH 66/680).
23 In 1875 and 1881 new sick wards were added at Crosland Moor and Deanhouse. Both institutions were later converted to geriatric hospitals and renamed St Luke's and St Mary's, respectively.
24 *Huddersfield Daily Chronicle*, 8 August 1872.
25 J. Davy, Workhouse report, 3 September 1885: MH 12/15111.
26 Crowther, *Workhouse system*, 58–9; K. Williams, *From pauperism to poverty* (London, 1981), 169–72.
27 CCE Minutes, PP 1849 XLII, 199.
28 *Leeds Mercury*, 4 March 1848. See also P. Pennock, 'The evolution of St James's', *Publications of the Thoresby Society* 59 (1987), 130–76.
29 A. Austin, Memorandum on R. Singer to PLB, 28 May 1849: MH 12/15071.
30 H. Farnall, Memorandum on C. Floyd to PLB, 14 April 1855: MH 12/15073.
31 *Huddersfield Chronicle*, 6 March 1858; J. Manwaring, Workhouse report, 31 May 1860, MH 12/15078.
32 J. Cocking to LGB, 2 January 1872: MH 12/15086.
33 J. Hall to LGB, 18 November 1880: MH 12/15098.
34 J. Hall to LGB, 25 May 1882: MH 12/15102.
35 W. Chance, *Children under the Poor Law* (London, 1897), 38. The Huddersfield

Guardians promoted the policy more widely: see *Poor Law Conferences* (1883), 528; *Poor Law Conferences* (1888), 67–8.

36 Jagger, *The history of Honley*, 326–7; J. Mozley, Report, 20 October 1894, MH 12/15129; E. Siddon, 'Women's work in the administration of the Poor Law', *Poor Law Conferences* (1908), 171–81. See also P. Hollis, *Ladies elect: women in English local government, 1865–1914* (Oxford, 1987), 195–299.

37 Annual Report of Insane Paupers, January 1857: MH 12/15074.

38 A Parliamentary return suggests that nearly two-thirds of adult paupers resident in Huddersfield Union workhouses for a continuous period of five years or more were classed as insane: PP 1861 LV.

39 Asylum bills constituted 5% of local relief expenditure in 1857; by 1885, the proportion had increased to 16%.

40 Commissioners in Lunacy, Workhouse report, March 1849: MH 12/15071.

41 J. Forster to PLB, 10 June 1857: MH 12/15075.

42 *Huddersfield Chronicle*, 16 January 1858.

43 J. Manwaring, Memorandum on C. Floyd to PLB, 10 June 1858, MH 12/15076; *Supplement to the Twelfth Annual Report of the Commissioners in Lunacy*, PP 1859.1 IX.

44 *Huddersfield Chronicle*, 6 June 1857.

45 C. Floyd to PLB, 23 May 1857: MH 12/15074.

46 R. Cane to PLB, 6 January 1868: MH 32/9.

47 C. Floyd to PLB, 16 February 1859: MH 12/15077.

48 Compare D. Ashforth, 'The Poor Law in Bradford, c. 1834–1871' (unpublished Ph.D. thesis, University of Bradford, 1979), 493–7; Boyson, 'The new Poor Law', 51.

49 The case of Charlotte Phillips deserves greater attention than I can give to it here. I hope to return to it in a future work.

50 HGM, 18 November 1892.

51 HGM, 22 August 1873, 16 October 1874. In 1878, for example, the Union received a grant of £1,745, representing about 40% of its expenditure on asylum bills: HGM, 20 December 1878.

52 M. Marland, *Medicine and society in Wakefield and Huddersfield, 1780–1870* (Cambridge, 1987); Driver, 'The English bastile', 411.

53 Model Lodging House regulations, HO 45/90177/9588; R. C. Pollution of Rivers, PP 1867 XXXIII, Q5378.

54 J. Pritchett to LGB, 22 January 1875: MH 12/15091.

55 Report on Infectious Hospitals, PP 1882 XXX.II, 21.

56 *Huddersfield Daily Chronicle*, 21 September 1878.

**Conclusion**

1 Anon., 'The English bastile', *Social Science Review* 3 (1865), 197.

2 G. Rhodes to LGB, 1 July 1876, MH 12/15092. There is an ironic twist to this tale. John Hall, the clerk to the Huddersfield Guardians in 1876, was later forced to resign after complaints about his drunkenness at Board meetings (J. Kilburn to LGB, 23 September 1896, MH 12/15131).

3 G. Procacci, 'Social economy and the government of poverty' in G. Burchell, C. Gordon and P. Miller (eds.), *The Foucault effect* (London, 1991), 161–2.

# Bibliography

I shall not weary you as I have wearied myself in endeavouring to obtain definite information on this subject. I could hardly devise a more cruel punishment for my worst enemy than to have to wade through the blue books of the Poor Law and Local Government Boards.

D. Tuke, *The past and present provision for the insane in Yorkshire* (London, 1889), p. 17.

## Manuscript sources

### *1 Private papers*


*Bentham Papers*: University College, London
*Brougham Papers*: University College, London
*Chadwick Papers*: University College, London
*Ellice Papers*: National Library of Scotland, Edinburgh
*Fielden Papers*: John Rylands Library, Manchester
*Harewood Papers*: Leeds Archives Department, Leeds
*Owen Papers*: Co-operative Union, Holyoake House, Manchester
*Place Papers*: British Library Manuscripts Division, London
*Russell Papers*: Public Records Office, London
*Starkey Papers*: Nottinghamshire Records Office, Nottingham
*Vincent–Minikin Papers*: Labour Party, London


### *2 Official records (Public Records Office)*


Central Poor Law authority minutes, MH 1
Registers of local Poor Law officers, MH 9
Poor Law Union papers and correspondence, MH 12
General Board of Health correspondence, MH 13
Central Poor Law authority correspondence, MH 19
Poor Law inspectors correspondence, MH 32
Registers of authorised workhouse expenditure, MH 34
Commissioners in Lunacy papers, MH 51

Surveys of health services in 1930s, MH 66
Poor Law authority establishment files, MH 78

Home Office disturbances papers, HO 40
Home Office disturbances entry books, HO 41
Home Office domestic entry books, HO 43
Home Office miscellaneous registered papers, HO 45
Home Office municipal and provincial correspondence, HO 52
Metropolitan Police correspondence, HO 61
Home Office correspondence with Poor Law authority, HO 73

Justices of the Peace, Oath books, C193
Justices of the Peace, Fiat books, C234

*3 Official records (Local government)*

A comprehensive list and description of all the local records consulted will be found in F. Driver, 'The English bastile' (unpublished Ph.D. thesis, 1987).

Honley township, poorhouse documents, 1763–7 (KDA)
Huddersfield borough, council minutes, 1868–1882 (KDA)
Huddersfield township, register of apprentices, 1798–1844 (TM)
Huddersfield township, town's meetings, 1835–1878 (KDA)
Huddersfield Union, Guardians minutes, 1837–1929 (KDA)
Huddersfield Union, Lists of paupers on relief, 1875–1913 (KDA)
Leeds Union, Moral and Industrial School plans (LA)
South Crosland township, overseers book, 1790–1801 (KDA)
Thurstonland township, town books, 1775–1843 (TM)

**Parliamentary papers**

*1 Continuous Series*

Annual Reports of Poor Law Commission (1835–1847)
Annual Reports of the Poor Law Board (1848–1871)
Annual Reports of the Local Government Board (1872–1886)
Minutes of the Committee of Council of Education (1849–1857)
Annual Returns of Relief Expenditure (1813–1886)
Annual Returns of Pauperism (1857–1886)

*2 Individual Reports*

Report on Continuance of the Poor Law Commission, PP 1840 XVII
Supplement to the Twelfth Annual Report of the Commissioners in Lunacy, PP 1859.1 IX
Report on Metropolitan Workhouses, PP 1866 LXI
Report on Workhouse Dietaries, PP 1866 XXXV
Reports on Vagrancy, PP 1866 XXXV

Report on Provincial Workhouses, PP 1867–8 LX
Reports on the Poor Law in Foreign Countries, PP 1875 LXV
Report on Cottage Homes, PP 1878 LX
Reports on Infectious Hospitals, PP 1882 XXX.II
Report of the Departmental Committee on Vagrancy, 1906 CIII

*3 Individual Returns*

Return of Relief Expenditure (1776), H.C. Committee Reports, IX
Return of Gilbert's Unions, PP 1844 XL
Return of Assistant Commissioners, PP 1845 XXXVIII
Return of Workhouse Accommodation, PP 1854–5 XLVI
Return of Gilbert's Unions and Local Act Unions, PP 1856 XLIX
Return of Workhouse Inmates, PP 1861 LV
Return of Workhouse Lunatic Wards, PP 1863 LII
Return of Cottage Homes, PP 1903 LIX

*4 Parliamentary Inquiries*

Royal Commission, Poor Laws, PP 1834 XXVII
Royal Commission, Poor Law (Official Reports), PP 1834 XVIII
Royal Commission, Poor Laws (Town Queries), PP 1834 XXXV
Royal Commission, Poor Laws (Miscellaneous), PP 1834 XXXVII
Royal Commission, Popular Education, PP 1861 XXI.I
Royal Commission, Pollution of Rivers, PP 1867 XXXIII
Royal Commission, Poor Laws (LGB Evidence), PP 1909 XXXIX
Royal Commission, Poor Laws (Northern Unions), PP 1909 XLI
Royal Commission, Poor Laws (Memoranda), PP 1910 LI

Select Committee, Manufactures, Commerce and Shipping, PP 1833 VI
Select Committee, Poor Laws, PP 1837–8 XVIII
Select Committee, Labouring Poor (Allotments), PP 1843 VII
Select Committee, Medical Relief, PP 1844 IX
Select Committee, Poor Relief (Gilbert Unions), PP 1844 X
Select Committee, Criminal & Destitute Juveniles, PP 1852 VII
Select Committee, Criminal & Destitute Children, PP 1852–3 XXIII
Select Committee, Care of Lunatics, PP 1859 III
Select Committee, Poor Relief, PP 1861 IX
Select Committee, Poor Relief, PP 1862 X

**Contemporary publications**

*1 Periodicals and newspapers*

*Contemporary Review*
*Edinburgh Review*
*Halifax Express*
*Halifax Guardian*

*Huddersfield Chronicle*
*Huddersfield and Holmfirth Examiner*
*Leeds Intelligencer*
*Leeds Mercury*
*Leeds Times*
*Macmillan's Magazine*
*Northern Star*
*Poor Law Conferences*
*The Times*
*Transactions of the National Association for the Promotion of Social Science*
*Westminster Review*

*2 Books and articles*

Aikin, J. *A description of the country from thirty to forty miles around Manchester* (London, 1795)
Anon., *An account of several workhouses for employing and maintaining the poor* (London, 1732)
*Great meeting ... of the factory children* (Leeds, 1833)
*Report of the sayings and doings at the grand yellow dinner* (Huddersfield, 1834)
*A report of the important proceedings of a public meeting held in the Chapel, Union Street, Oldham* (Oldham, 1836)
*Give it a fair trial* (Huddersfield, c. 1837)
'Schools of industry', *Chambers' Miscellany* 14 (1846), 1–32
'Patronage of Commissions', *Westminster Review* 46 (1847), 222–45
'The English Bastile', *Social Science Review* 3 (1865), 193–200
'Pauper girls', *Westminster Review*, 37 (1870), 461–76
'Sir Alfred Power, KCB', *New Monthly Magazine* 117 (1880), 390–4
'The work of women as Poor Law Guardians', *Westminster Review* 67 (1885), 386–95
Anstie, F., 'Insane patients in London workhouses', *Journal of Mental Science* 11 (1865), 327–36
Archbold, J., *The Consolidated and other Orders of the Poor Law Commissioners and the Poor Law Board* (London, 1859)
Austin, J., 'Centralisation', *Edinburgh Review* 85 (1847), 221–58
Becher, J. T., *The anti-pauper system* (London, 1828)
Bowring, J., (ed.), *The works of Jeremy Bentham* (London, 1843)
Brodrick, G., 'Local government in England' in J. W. Probyn (ed.), *Local government and taxation* (London, 1875), 1–95
Bucknill, J. C., 'The custody of the insane poor', *Asylum Journal* 25 (1858), 460–72
Butler, J. E., (ed.), *Woman's work and woman's culture* (London, 1869)
Carpenter, M., 'On the education of pauper girls', *Transactions, NAPSS* (1862), 240–73
Chadwick, E., 'The new Poor Law', *Edinburgh Review* 63 (1836), 487–537
Chance, W., *Children under the Poor Law: their education, training and after-care* (London, 1897)
Cobbe, F. P., 'The philosophy of the Poor Laws', *Fraser's Magazine* 70 (1864), 373–94

Copleston, E., *A second letter to the Right Honourable Robert Peel ... on the cause of the increase in pauperism and on the Poor Laws* (Oxford, 1819)

Craven, C., *A night in the workhouse* (Keighley, 1887)

Dicey, A. V., *Lectures on the relationship between law and public opinion in England during the nineteenth century* (London, 1905)

Doyle, A., *Proposed district school on the system of Mettray* (London, 1873)

Duncombe, E., *Gilbertise the new Poor Law: a fresh plan instead of the new Poor Law, police and fiscal laws* (York, 1841)

Eden, F. M., *The state of the poor: an history of the labouring classes in England from the conquest to the present period* (London, 1797)

Farrington, J., *Jim o'th' Pan's journey to London, with the new Poor Law to mend, by a Collector* (Huddersfield, 1842)

Ferguson, S., 'Architecture in Ireland', *Dublin University Magazine* 29 (1847), 693–708

Fletcher, J., 'Statistics of the farm school system on the continent', *Journal of the Statistical Society of London* 15 (1852), 1–49

Fletcher, M., *Migration of agricultural labourers* (Bury, 1837)

Foster, J., *Pedigrees of the county families of Yorkshire* (London, 1874)

Freeman, A., *Hints on the planning of Poor Law buildings* (London, 1904)

Gilbert, T., *Considerations on the Bills for the better relief and employment of the poor* (London, 1787)

Gilbert Scott, G., *Personal and professional recollections* (London, 1879)

Hall, J., 'The proper employment of pauper labour', *Poor Law Conferences* (1883), 519–29

Hanson, J., *View extraordinary of Sir John's Huddersfield menagerie* (Leeds, 1837)

Herford, C. J., *The education of orphan pauper children* (Manchester, 1865)

Hill, F. D., 'The family system for workhouse children', *Contemporary Review* 15 (1870), 240–73

*Children of the state* (London, 1868; second edn, 1889)

'The system of boarding out pauper children', *Economic Journal* 3 (1893), 62–73

Hole, J., *Lectures on social science and the organisation of labor* (London, 1851)

Hovell, M., *The Chartist movement* (London, 1925)

I'Anson, B., *The history of the Armytage or Armitage family* (London, c. 1915)

Jagger, M., *The history of Honley* (Huddersfield, 1914)

Leigh, J., 'Juvenile offenders and destitute pauper children' in Viscount Ingestre (ed.), *Meliora* vol. II (London, 1853), 81–9

Lindsay, W. L., 'The family system as applied to the treatment of the chronic insane', *Journal of Mental Science* 16 (1871), 497–527

Locke, J., *Report of the Board of Trade to the Lords Justices in the year 1697* (reprinted: London, 1789)

Malthus, T. R., *Essay on population*, 6th edn (London 1826)

Maunsell, H., *Political medicine* (Dublin, 1839)

Monnington, M. W. and Lampard, F. J., *Our Poor Law schools* (London, 1898)

Murray-Browne, T. L., 'The education and future of workhouse children', *Poor Law Conferences* (London, 1883), 93–8

Napier, W., *The life and opinions of General Sir Charles James Napier* (London, 1857)

Nicholls, G., *A history of the English Poor Law* (London, 1854; repr. 1898)

Oastler, R., *Speech delivered at a public meeting held in the Market place, Huddersfield* (Leeds, 1833)
*Eight letters to the Duke of Wellington* (London, 1835)
*A letter to the Bishop of Exeter* (Manchester, 1838)
*The right of the poor to liberty and life* (London, 1838)
Pashley, R., *Pauperism and Poor Laws* (London, 1852)
Phillips, G. S., *Walks around Huddersfield* (Huddersfield, 1848)
Preston-Thomas, H., *The work and play of a government inspector* (Edinburgh, 1909)
Prichard, J., *A treatise on insanity* (London, 1835)
Pugin, A. W. N., *Contrasts*, second edn (London, 1841)
Redlich, J. and Hirst, F., *Local government in England* (London, 1903)
Rhodes, G., 'Classification of paupers by workhouses', *Poor Law Conferences* (1895), 266–77
Roberts, S., *The peers, the people and the poor* (London, 1838)
Robertson, C. L., 'The care and treatment of the insane poor', *Journal of Mental Science* 13 (1867), 289–306
Russell, R. (ed.), *The early correspondence of Lord John Russell, 1805–1840* (London, 1913)
Sadler, M. T., *Protest against the secret proceedings of the Factory Commission* (Leeds, 1833)
Senior, N., *Three lectures on the rate of wages . . . with a preface on the causes and remedies of the present disturbances* (London, 1830)
*Remarks on the opposition to the Poor Law Amendment Bill* (London, 1841)
'English Poor Laws', (orig. 1841) in *Historical and philosophical essays* (London, 1865), vol. II, 45–115
Siddon, E., 'Women's work in the administration of the Poor Law', *Poor Law Conferences* (1908), 171–81
Smith, P. G., *Hints and suggestions as to the planning of Poor Law buildings* (London, 1904)
Stallard, J. H., *Workhouse hospitals* (London, 1865)
Sykes, J., *Slawit in the sixties* (Huddersfield, 1926)
Symons, J. C., 'On industrial training' in A. Hill (ed.), *Essays on educational subjects* (London, 1857), 37–56
Synnot, H., 'Little paupers', *Contemporary Review* 24 (1874), 954–72
Tatham, T. R., *Medical relief: Mr Tatham's case against the Huddersfield Guardians* (Huddersfield, 1848)
Titmarsh, M. A. [W. Thackeray], *The Irish sketch-book* (London, 1843)
Tocqueville, A. de, *Correspondence and conversations with Nassau Senior* (London, 1872)
Tuke, D. H., *The past and present provision for the insane in Yorkshire* (London, 1889)
Tuke, S., 'Practical hints on the construction of pauper lunatic asylums' in *Plans of the pauper lunatic asylum lately erected at Wakefield* (York, 1819), 5–23
Turner, S., *Reformatory schools* (London, 1855)
Turner, S. and Paynter, T., *Report on the system and objects of la colonie agricole at Mettray* (London, 1846)
Twining, L., 'Workhouse education', *Transactions, NAPSS* (1861), 331–8
'Women as official inspectors', *Nineteenth Century* 35 (1894), 489–94

Wade, J., *Appendix to the Black Book* (London, 1835)
Walker, J. K., 'A sketch of the medical and general topography of Huddersfield and its adjoining district', *London Medical Repository* 10 (1818), 1–16
Wythen Baxter, G., *The book of the bastiles* (London, 1841)
Zouch, H., *Remarks upon the late resolutions of the House of Commons respecting the proposed change of the Poor Laws* (Leeds, 1776)

## Secondary sources

### 1 Unpublished theses

Ashforth, D., 'The Poor Law in Bradford, 1834–1871' (Ph.D. thesis, University of Bradford, 1979)
Corrigan, P. 'State formation and moral regulation in nineteenth-century Britain' (Ph.D. thesis, University of Durham, 1977)
Driver, F. 'The English bastile: dimensions of the workhouse system, 1834–1884' (Ph.D. thesis, University of Cambridge, 1987)
Peacock, A. J., 'The Justices of the Peace and the prosecution of the Factory Acts' (Ph.D. thesis, University of York, 1982)
Pierce, E. M., 'Town–country relations in England and Wales in the pre-railway age, as revealed by the Poor Law Unions' (M.A. thesis, University of London, 1957)
Rose, M. E., 'Poor Law administration in the West Riding of Yorkshire (D.Phil. thesis, University of Oxford, 1965)

### 2 Books and articles

Alliès, P., *L'invention du territoire* (Grenoble, 1980)
Anderson, P., 'The Leeds workhouse under the old Poor Law, 1726–1834', *Publications of the Thoresby Society* 56 (1980), 75–113
Annette, J., 'Bentham's fear of hobgoblins' in B. Fine, R. Kinsey, J. Lea, S. Picciotto and J. Young (eds.), *Capitalism and the rule of law* (London, 1979), 65–75
Ashforth, D., 'The urban Poor Law' in D. Fraser (ed.), *The new Poor Law in the nineteenth century* (London, 1976), 128–48
Ayers, G. M., *England's first state hospitals and the Metropolitan Asylums Board* (London, 1971)
Bahmueller. C. F., *The national charity company: Jeremy Bentham's silent revolution* (Berkeley, 1981)
Bartrip, P., 'British government inspection, 1832–1875', *Historical Journal* 25 (1982), 605–26
'State intervention in nineteenth-century Britain: fact or fiction?', *Journal of British Studies* 23 (1983), 63–83
'Success or failure? The prosecution of the early Factory Acts', *Economic History Review* 38 (1985), 423–7
Bédarida, F., 'L'Angleterre Victorienne: paradigme du *laissez-faire*', *Revue Historique* 261 (1979), 79–98
Bellamy, C., *Administering central–local relations, 1871–1919: the Local Government Board in its fiscal and cultural context* (Manchester, 1988)

Blaug, M., 'The myth of the old Poor Law and the making of the new', *Journal of Economic History* 23 (1963), 151–84

Bowley, M., *Nassau Senior and classical economics* (London, 1937)

Boyer, G. R., *An economic history of the English Poor Law, 1750–1850* (Cambridge, 1990)

Boyson, R., 'The new Poor Law in North-east Lancashire', *Transactions, Lancashire and Cheshire Antiquarian Society* (1960), 35–56

Brewer, J., *The sinews of power: war, money and the English state, 1688–1783* (London, 1989)

Briggs, A., 'The welfare state in historical perspective', *Archives of European Sociology* 2 (1961), 221–58

Brundage, A., 'The landed interest and the new Poor Law', *English Historical Review* 87 (1972), 27–48

'The new Poor Law and the cohesion of agricultural society', *Agricultural History* 47 (1974), 405–17

*The making of the new Poor Law, 1832–1839* (London, 1978)

Bulmer, M., Lewis, J. and Piachaud, D. (eds.), *The goals of social policy* (London, 1989)

Burchell, G., Gordon, C. and Miller, P. (eds.), *The Foucault effect: studies in governmentality* (London, 1991)

Cannadine, D., *Land and landlords: the aristocracy and the towns, 1774–1967* (Leicester, 1980)

Caplan, M., 'The new Poor Law and the struggle for Union chargeability', *International Review of Social History* 23 (1978), 267–300

Carlson, E. and Dain, N., 'The meaning of moral insanity', *Bulletin of the History of Medicine* 36 (1962), 130–40

Carson, W., 'Symbolic and instrumental dimensions of early factory legislation' in R. Hood (ed.), *Crime, criminology and public policy* (London, 1974), 107–38

Checkland, S. and Checkland, E., *The 1834 Poor Law Report* (Harmondsworth, 1974)

Claeys, G., *Citizens and saints: politics and anti-politics in early British socialism* (Cambridge, 1989)

Clokie, H. and Robinson, J., *Royal Commissions of inquiry: the significance of investigations in British politics* (Stanford University, 1937)

Coats, A. W., 'Economic thought and Poor Law policy in the eighteenth century', *Economic History Review* 13 (1960–1), 39–51

(ed.), *The classical economists and economic policy* (London, 1971)

Cohen, S., 'The punitive city: notes on the dispersal of social control', *Contemporary Crises* 3 (1979), 339–63

*Visions of social control* (Cambridge, 1985)

Cole, D., 'Sir George Gilbert Scott' in P. Ferriday (ed.), *Victorian architecture* (London, 1963), 175–84

Cole, J. (ed.), *Down workhouse lane: the diary of a Rochdale workhouse* (Littleborough, 1984)

Collini, S., *Public moralists: political life and intellectual thought in Britain, 1850–1930* (Oxford, 1991)

Corrigan, P., 'On moral regulation' in P. Corrigan, *Social forms/human capacities: essays in authority and difference* (London, 1990), 102–29

Corrigan, P. and Sayer, D., *The great arch* (Oxford, 1985)
Cosgrove, R. A., *The rule of law: A. V. Dicey, Victorian jurist* (London, 1980)
Crowther, M. A., *The workhouse system, 1834–1929* (London, 1981)
Crump, W. and Ghorbal, G., *History of the Huddersfield woollen industry* (Huddersfield, 1935)
Cullen, M., *The statistical movement in early Victorian Britain* (Brighton, 1975)
Dandeker, C., *Surveillance, power and modernity: bureaucracy and discipline from 1700 to the present day* (Cambridge, 1990)
Daniels, S. and Seymour, S., 'Landscape design and the idea of improvement, 1730–1900' in R. Dodgshon and R. Butlin (eds.), *An historical geography of England and Wales* (London, 1990), 487–520
Dean, M., *The constitution of poverty: toward a genealogy of liberal governance* (London, 1991)
Dear, M. and Wolch, J., *Landscapes of despair: from deinstitutionalisation to homelessness* (Cambridge, 1987)
Dickens, A., 'The architect and the workhouse', *Architectural Review* 160 (1976), 345–52
Digby, A., *Pauper palaces* (London, 1978)
'The principles of 1834', *Middlesbrough Centre, Occasional Paper* 1 (1985), 1–14
Donajgrodski, A., 'Social police and the bureaucratic elite' in A. Donajgrodski (ed.), *Social control in nineteenth-century Britain* (London, 1976), 51–76
Donnelly, M., *Managing the mind: a study of medical psychology in early nineteenth-century Britain* (London, 1983)
Driver, C., *Tory Radical: the life of Richard Oastler* (New York, 1946)
Driver, F., 'Power, space and the body: a critical assessment of Foucault's *Discipline and Punish*', *Society and Space* 3 (1985), 425–46
'Moral geographies: social science and the urban environment in mid-nineteenth century England', *Transactions, Institute of British Geographers* 13 (1988), 275–87
'The historical geography of the workhouse system, 1834–1883', *Journal of Historical Geography* 15 (1989), 269–86
'Discipline without frontiers? Representations of the Mettray reformatory colony in Britain, 1840–1880', *Journal of Historical Sociology* 3 (1990), 272–93
'Tory Radicalism? Ideology, strategy and locality in popular politics during the 1830s', *Northern History* 27 (1991), 120–38
'Bodies in space: Foucault's account of disciplinary power' in C. Jones and R. Porter (eds.) *Reassessing Foucault* (London, 1993)
Dunkley, P., 'The landed interest and the making of the new Poor Law: a critical note', *English Historical Review* 88 (1973), 836–41
'The hungry forties and the new Poor Law: a case study', *Historical Journal* 2 (1974), 329–46
*The crisis of the old Poor Law, 1795–1834* (New York, 1982)
Dunkley, P. and Apfel, W., 'English rural society and the new Poor Law', *Social History* 10 (1985), 37–68
Dworkin, R., *Taking rights seriously* (London, 1977)
Dyson, K., *The state tradition in Western Europe* (Oxford, 1980)
Edsall, N. C., *The anti-Poor Law movement, 1834–1844* (Manchester, 1971)

Epstein, J., *The lion of freedom: Feargus O'Connor and the Chartist movement, 1832–1842* (London, 1983)
Evans, R., 'Bentham's Panopticon' *Architectural Association Quarterly*, July 1971, 21–37
*The fabrication of virtue: English prison architecture, 1750–1840* (Cambridge, 1982)
Finer, S. E., *The life and times of Sir Edwin Chadwick* (London, 1952)
'The transmission of Benthamite ideas, 1820–1850' in G. Sutherland (ed.), *Studies in the growth of nineteenth-century government* (London, 1972), 11–32
Finnane, M., 'Asylums, families and the state', *History Workshop Journal* 20 (1985), 134–48
Foster, J., *Class struggle and the industrial revolution* (London, 1974)
Foucault, M., *Madness and civilisation: a history of insanity in the age of reason* (London, 1967)
*Discipline and punish: the birth of the prison* (Harmondsworth, 1977)
*The history of sexuality*, Vol. I (London, 1979)
'Governmentality', *Ideology and Consciousness* 6 (1979), 9–21
'Space, knowledge and power', *Skyline*, March 1982, 16–20
Foucault, M., Barret-Kriegel, B., Thalamy, A., Béguin, F. and Fortier, B., *Les machines à guérir: aux origines de l'hôpital moderne* (Brussels, 1976)
Fraser, D., 'Poor Law politics in Leeds', *Publications of the Thoresby Society* 53 (1970), 23–49
'The Poor Law as a political institution' in D. Fraser (ed.), *The new Poor Law in the nineteenth century* (London, 1976), 111–27
*The evolution of the British welfare state*, second edn (London, 1984)
Garrard, J., *Leadership and power in Victorian industrial towns, 1830–1850* (Manchester, 1983)
Gash, N., *Reaction and reconstruction in English politics, 1832–1852* (Oxford, 1965)
Gatrell, V., 'Crime, authority and the policeman-state' in F. M. L. Thompson (ed.), *The Cambridge social history of Britain*, vol. III (Cambridge, 1990), 243–310
Giddens, A., *A contemporary critique of historical materialism*, Vol. II: *The nation-state and violence* (Cambridge, 1985)
Ginsburg, N., *Class, capital and social policy* (London, 1979)
Goffman, E., *Asylums* (Harmondsworth, 1968)
Goldthorpe, J. H., 'The development of social policy in England 1800–1914', *Transactions, Fifth World Congress of Sociology* 4 (1962), 41–56
Gordon, C., 'Afterword' in Gordon, C. (ed.), *Power/Knowledge: selected interviews and other writings by Michel Foucault* (Brighton, 1980), 229–59
Gough, I., *The political economy of the welfare state* (London, 1979)
Greenleaf, W., 'Toulmin Smith and the British political tradition', *Public Administration* 53 (1975), 25–44
Grew, R., 'The nineteenth-century European state' in C. Bright and S. Harding (eds.), *State-making and social movements* (Ann Arbor, 1984), 83–120
Hacking, I., 'How should we do the history of statistics?' in G. Burchell, C. Gordon and P. Miller (eds.), *The Foucault effect: studies in governmentality* (London, 1991), 181–95
Halévy, E., *The growth of philosophic radicalism* (London, 1928)
Hall, J. (ed.), *States in history* (Oxford, 1986)

Harris, C., 'Power, modernity and historical geography', *Annals, Association of American Geographers* 81 (1991), 671–83

Harris, J., *British government inspection: local services and the central departments* (London, 1955)

Harrison, J. F. C., *Robert Owen and the Owenites in Britain and America* (London, 1969)

Hart, J., 'Nineteenth-century social reform: a Tory interpretation of history', *Past and Present* 31 (1965), 39–61

Heidenheimer, A., 'Politics, policy and *policey* as concepts in English and continental languages', *Review of Politics* 48 (1986), 3–30

Hennock, E. P., 'Central/local relations in England: an outline', *Urban History Yearbook* (1982), 38–49

Henriques, U., 'Jeremy Bentham and the machinery of social reform' in H. Hearder and H. Loyn (eds.), *British government and administration* (Cardiff, 1974), 169–86

Hewitt, M., 'Social policy and the politics of life: Foucault's account of welfare', *Occasional Paper, School of Social Sciences, Hatfield Polytechnic* 7 (1982)

Hobsbawm, E. and Rudé, G., *Captain Swing* (New York, 1968)

Hodgkinson, R., *The origins of the National Health Service: the medical services of the new Poor Law, 1834–1871* (London, 1967)

Hollis, P., *Ladies elect: women in English local government, 1865–1914* (Oxford, 1987)

Hume, L. J., 'Jeremy Bentham and the nineteenth-century revolution in government', *Historical Journal* 10 (1967), 361–75

*Bentham and bureaucracy* (Cambridge, 1981)

Ignatieff, M., *A just measure of pain: the penitentiary in the industrial revolution, 1750–1850* (London, 1978)

'State, civil society and total institutions: a critique of recent social histories of punishment', *Crime and Justice* 3 (1981), 153–92

Jackman, S., *Galloping Head: the life of the Right Honourable Sir Francis Bond Head, 1793–1875* (London, 1958)

Johnson, R., 'Educating the educators: "experts" and the state, 1833–1839' in A. Donajgrodski (ed.), *Social control in nineteenth-century Britain* (London, 1977), 77–107

Jones, D., *Crime, protest, community and police in nineteenth-century Britain* (London, 1982)

Kearns, G., 'Private property and public health reform in England, 1830–1870', *Social Science and Medicine* 26 (1988), 187–99

Kearns, G. and Withers, C. (eds.), *Urbanising Britain: essays on class and community in the nineteenth-century city* (Cambridge, 1991)

King, A. D. (ed.), *Buildings and society: essays on the social development of the built environment* (London, 1980)

Kitson Clark, G., 'Statesmen in disguise', *Historical Journal* 2 (1959), 19–39

Knott, J., *Popular opposition to the 1834 Poor Law* (London, 1986)

Lambert, R., *Sir John Simon, 1816–1904, and English social administration* (London, 1963)

Lipman, V., *Local government areas, 1834–1945* (Oxford, 1949)

Lowman, J., 'Conceptual issues in the geography of crime: toward a geography of social control', *Annals, Association of American Geographers* 76 (1986), 81–94

Luckin, B., 'Towards a social history of institutionalization', *Social History* 8 (1983), 87–94

Macdonagh, O., 'The nineteenth-century revolution in government: a reappraisal', *Historical Journal* 1 (1958), 52–67

*Early Victorian government* (London, 1977)

Mackinnon, M., 'English Poor Law policy and the crusade against out-relief', *Journal of Economic History* 47 (1987), 603–25

MacLeod, R., (ed.), *Government and expertise: specialists, administrators and professionals, 1860–1919* (Cambridge, 1988)

Maiwald, K., 'An index of building costs in the United Kingdom, 1845–1938', *Economic History Review* 7 (1954), 187–203

Mandler, P., 'The making of the new Poor Law *redivivus*', *Past and Present* 117 (1987), 131–57

Mann, M., 'The autonomous power of the state', *Archives of European Sociology* 25 (1984), 185–213

*The sources of social power*, vol. I (Cambridge, 1986)

Markus, T. (ed.), *Order in space and society: architectural form and its context in the Scottish Enlightenment* (Edinburgh, 1982)

Marland, M. H., *Medicine and society in Wakefield and Huddersfield, 1780–1870* (Cambridge, 1987)

Marshall, D., 'The role of the Justice of the Peace' in H. Hearder and H. Loyn (eds.), *British government and administration* (Cardiff, 1974), 155–68

Marshall, J. D., *The old Poor Law, 1795–1834* (London, 1968)

Martin, E. W. 'From parish to Union' in E. W. Martin (ed.), *Comparative development in social welfare* (London, 1972), 25–56

Mather, F., *Public order in the age of the Chartists* (Manchester, 1959)

McCandless, P., 'Build! Build! The controversy over the care of the chronically insane in England, 1835–1870', *Bulletin for the History of Medicine* 53 (1979), 553–74

Mellett, D. J., *The prerogative of asylumdom* (New York, 1982)

Melossi, D. and Pavarini, M., *The prison and the factory* (London, 1981)

Midwinter, E. C., 'State intervention at the local level: the new Poor Law in Lancashire', *Historical Journal* 10 (1967), 106–12

*Social administration in Lancashire, 1830–1860* (Manchester, 1969)

Mort, F., *Dangerous sexualities: medico-moral politics in England since 1830* (London, 1987)

Novak, T., *Poverty and the state: an historical sociology* (Milton Keynes, 1988)

O'Brien, D. P., *The classical economists* (Oxford, 1975)

Ogborn, M., 'A lynx-eyed and iron-handed system: the state regulation of prostitution in nineteenth-century Britain' (unpublished conference paper, 1989)

'Local power and state regulation in nineteenth-century Britain', *Transactions, Institute of British Geographers* 17 (1992), 215–26

Oxley, G. W., 'The permanent poor in South-West Lancashire under the old Poor Law' in J. Harris (ed.), *Liverpool and Merseyside* (Liverpool, 1969), 16–49

Paddison, R., *The fragmented state: the political geography of power* (Oxford, 1983)

Parris, H., 'The nineteenth-century revolution in government: a reappraisal reappraised', *Historical Journal* 3 (1960), 17–37

Parry-Jones, W. L., 'The model of the Geel lunatic colony and its influence on the nineteenth-century asylum system in Britain' in A. Scull (ed.), *Madhouses, mad-doctors and madmen* (London, 1981), 201–17

Pasquino, P., 'Theatrum politicum', *Ideology and Consciousness* 4 (1978), 41–54

Paterson, F., *Out of place: public policy and the emergence of truancy* (London, 1989)

Peake, H., 'Geographical aspects of administrative areas', *Geography* 15 (1930), 531–46

Pennock, P., 'The evolution of St James's, Leeds Moral and Industrial Training School, Leeds Union Workhouse and Leeds Union Infirmary', *Thoresby Society Publications* 59 (1987), 130–76

Perkin, H., 'Individualism versus collectivism in nineteenth-century Britain: a false antithesis', *Journal of British Studies* 17 (1977), 105–18

Perrot, M. (ed.), *L'impossible prison: recherches sur le système penitentaire au XIXe siècle* (Paris, 1980)

Perry, P. J., *Geography of nineteenth-century Britain* (London, 1975)

Philips, D., 'A new engine of power and authority' in G. Parker, B. Lenman and V. Gatrell (eds.), *Crime and the law: the social history of crime in Western Europe since 1500* (London, 1980), 155–89

Philo, C., 'Fit localities for an asylum', *Journal of Historical Geography* 13 (1987), 398–415

'Enough to drive one mad: the organisation of space in nineteenth-century lunatic asylums' in J. Wolch and M. Dear (eds.), *The power of geography* (London, 1989), 258–90

Pickering, P. A., 'Class without words', *Past and Present* 112 (1986), 144–62

Piven, F. and Cloward, R., *Regulating the poor: the functions of public welfare* (New York, 1971)

Polanyi, K., *The great transformation* (New York, 1945)

'Our obsolete market mentality' in G. Dalton (ed.), *Primitive, archaic and modern economies* (New York, 1968), 59–77

Porter, R., *Mind-forg'd manacles: a history of madness in England from the Restoration to the Regency* (London, 1987)

Poynter, F., *Society and pauperism: English ideas on poor relief, 1795–1834* (London, 1969)

Prest, J., *Liberty and locality: Parliament, permissive legislation and ratepayers' democracies in the nineteenth century* (Oxford, 1990)

Procacci, G., 'Social economy and the government of poverty' in G. Burchell, C. Gordon and P. Miller (eds.), *The Foucault effect: studies in governmentality* (London, 1991), 151–68

Raeff, M., 'The well-ordered police state', *American Historical Review* 80 (1975), 1221–43

Raffestin, C., *Pour une géographie du pouvoir* (Paris, 1980)

Rendall, J., *The origins of modern feminism* (London, 1985)

Richards, P., 'The state and early industrial capitalism: the case of the hand-loom weavers', *Past and Present* 83 (1979), 91–115

'State formation and class struggle, 1832–1848' in P. Corrigan (ed.), *Capitalism, state formation and Marxist theory* (London, 1980), 49–78

Robbins, L., *The theory of economic policy in English classical political economy* (London, 1952)

Roberts, D., *Victorian origins of the British welfare state* (New Haven, 1960)
Robinson, J., 'A perfect system of control? State power and "native locations" in South Africa', *Environment and Planning D: Society and Space* 8 (1990), 135–62
Rose, M. E., 'The anti-Poor Law agitation' in J. T. Ward (ed.), *Popular movements, 1830–1850* (London, 1970), 78–94
'The new Poor Law in an industrial area' in R. Hartwell (ed.), *The industrial revolution* (Oxford, 1970), 121–43
'The crisis of poor relief in England, 1860–1914' in W. J. Mommsen (ed.), *The emergence of the welfare state in Britain and Germany* (London, 1981), 50–70
Rose, N., *The psychological complex: psychology, politics and society in England, 1869–1939* (London, 1985)
'Calculable minds and manageable individuals', *History of the Human Sciences* 1 (1988), 179–200
Rosen, G., 'Cameralism and the concept of medical police' in G. Rosen (ed.), *From medical police to social medicine* (New York, 1974), 142–58
Rosenberg, C., 'Florence Nightingale on contagion: the hospital as moral universe' in C. Rosenberg (ed.), *Healing and history* (New York, 1979), 116–36
Rothman, D., *The discovery of the asylum* (Toronto, 1971)
Sack, R., *Human territoriality* (Cambridge, 1986)
Schmiechen, J., 'The Victorians, the historians and the idea of modernism', *American Historical Review* 93 (1988), 287–314
Scull, A., 'The domestication of madness', *Medical History* 27 (1983), 233–48
*Decarceration: community treatment and the deviant* (Cambridge, 1984)
*Social order/mental disorder: Anglo-American psychiatry in historical perspective* (Berkeley, 1989)
Silver, A., 'The demand for order in civil society' in D. Bordua (ed.), *The police* (New York, 1967), 1–24
Simmel, G., 'The poor' (orig. 1908), *Social Problems* 13 (1965–6), 119–40
Smith, C., *Public problems: the management of urban distress* (New York, 1988)
Snell, K., *Annals of the labouring poor: social change and agrarian England, 1600–1900* (Cambridge, 1985)
Springett, J., 'Landowners and urban development: the Ramsden estate and nineteenth-century Huddersfield', *Journal of Historical Geography* 8 (1982), 129–44
Stedman Jones, G., *Outcast London* (Oxford, 1971)
'The language of Chartism' in J. Epstein and D. Thompson (eds.), *The Chartist experience* (London, 1982), 3–58
Stokes, H., 'Cambridge parish workhouses', *Proceedings of the Cambridge Antiquarian Society* 59 (1911), 70–132
Storch, R., 'The plague of blue locusts: police reform and popular resistance in Northern England, 1840–1857', *International Review of Social History* 20 (1975), 61–90
Taylor. J. S., 'The unreformed workhouse' in E. W. Martin (ed.), *Comparative development in social welfare* (London, 1972), 57–84
Thompson, D., *The Chartists: popular politics in the industrial revolution* (London, 1983)
Thompson, J. and Goldin, G., *The hospital: a social and architectural history* (New Haven, 1975)

Thompson, R., 'The working of the Poor Law Amendment Act in Cumbria, 1836–1871', *Northern History* (1981), 117–37

Thomson, D., 'Welfare and the historians' in L. Bonfield, R. Smith and K. Wrightson (eds.), *The world we have gained* (Oxford, 1986), 355–78

Viner, J., 'Adam Smith and *laissez-faire*' in J. Viner (ed.), *The long view and the short: studies in economic theory and policy* (Glencoe, Illinois, 1958), 213–45

Ward, D., *Poverty, ethnicity and the American city, 1840–1925* (Cambridge, 1989)

Weaver, S. A., *John Fielden and the politics of popular radicalism, 1832–1847* (Oxford, 1987)

Webb, B. and Webb, S., *Statutory authorities for specific purposes* (London, 1922)

*English Poor Law History*, Part I: *The old Poor Law* (London, 1927)

*English Poor Law History*, Part II: *The last hundred years* (London, 1929)

Wells, R., 'Resistance to the new Poor Law in the rural South', *Middlesbrough Centre, Occasional Paper* 1 (1985), 15–53

White, G. (ed.), *In and out of the workhouse: the coming of the new Poor Law to Cambridgeshire and Huntingdonshire* (Ely, 1978)

Wiener, M., *Reconstructing the criminal: culture, law and policy in England, 1830–1914* (Cambridge, 1990)

Williams, K., *From pauperism to poverty* (London, 1981)

Wolfe, D. A., 'Mercantilism, Liberalism and Keynesianism', *Canadian Journal of Social and Political Theory* 5 (1981), 69–96

Wood, P., 'Finance and the urban Poor Law: Sunderland Union, 1836–1914', in M. E. Rose (ed.), *The poor and the city* (Leicester, 1985), 20–56

Yeo, E., 'Culture and constraint in working-class movements, 1830–1855', in E. Yeo and S. Yeo (eds.), *Popular culture and class conflict* (Brighton, 1981), 155–86

Young, G. M., *Victorian England: portrait of an age* (London, 1936)

Young, R. M., 'Association of ideas' in P. Wiener (ed.), *Dictionary of the history of ideas* (London 1973), 111–18

# Index